Safety WALK
Safety TALK

Safety WALK Safety TALK

How small changes in what you
THINK, SAY, & DO
shape your safety culture

David Allan Galloway

www.ContinuousMILE.com.

Cover page photo attribution:
https://www.dreamstime.com/royalty-free-stock-image-work-helmet-image16324056#res15971830

Continuous MILE Consulting, LLC
Kindle Direct Publishing
Published January, 2019
Revised October, 2019
ISBN 13: 978-1729608722

To my wife, Leesa
and to our children, Rebecca and Rachel

"The purpose of life is not to be happy. It is to be useful, to be honorable, to be compassionate, to have it make some difference that you have lived and lived well."

- Ralph Waldo Emerson

Contents

Preface

Do you want to make a difference?

There are many ways someone in a leadership role can have a positive impact on the lives of their employees. Perhaps there is no leadership responsibility more profound than creating a sustainable, injury-free workplace. Every person who goes to work expects to return home in the same condition. When someone is hurt, the adverse effects of their injury ripple through the employee's family and friends.

Achieving an injury-free environment is one of the most difficult problems many leaders face. Indeed, during 35 years in manufacturing I never discovered a singular solution to this challenge. However, over those years I observed quite a few leadership actions that significantly contributed to less risk-taking, greater hazard awareness and genuine collaborative efforts among employees and supervisors. Leaders who understood, embraced, and implemented these strategies saw a dramatic reduction in incidents and injuries at their facilities.

Here's the thing. In my experience, organizations that exhibit the best safety performances do not have a secret unknown to poorer performing organizations. They simply do a lot of small things collectively and strategically. And they do them well.

That's really what this book is about. It is a collection of leadership concepts, thoughts, words, and actions that (when strategically implemented) can move your organization toward a better safety future. There are no 'silver bullets' here. On the other hand, you don't have to do all of these things to be successful in your safety journey.

Here is how the book is organized.

Part One takes a look at some fundamental concepts everyone who is striving to achieve safety excellence should understand. It includes a discussion on compliance versus commitment, how to develop a safety strategy, why people make mistakes and take risks, and an overview of a Just Culture.

Part Two through Part Six leverages research in social psychology, sociology and neuroscience, as well as real-life events and personal experiences to examine how these concepts can be used to improve your safety leadership skills. This is followed by a discussion about the practical application of what is presented. There is a segment at the end of each chapter in this section of the book called the **SAFETY LEADER'S TOOLBOX™**. This toolbox contains over 70 practical tools and tips for being a more effective safety leader!

Look for this icon:

Part Seven is essentially a "call to action" for leaders.

I invite you to put on your safety shoes and walk with me. Together we will consider how you can lead your organization to exceptional safety performance. Spoiler alert! One essential leadership skill is knowing *why, how,* and *what* to talk about when it comes to safety.

Where do you begin? If you start with caring as your primary motive, you won't have to do everything perfectly. Your employees will want to do the right things for the right reasons. They will eagerly join you on the safety journey.

You can read this book in chapter order. You can also go to a specific chapter to learn more about a particular topic. Either way, you are encouraged to consult the **SAFETY LEADER'S TOOLBOX™** throughout this book for small changes in what you think, say, and do to shape your safety culture.

Choose a set of leadership tools that will enable you to move toward your safety vision. Align your team around these actions.

Start making a difference in the lives of others!

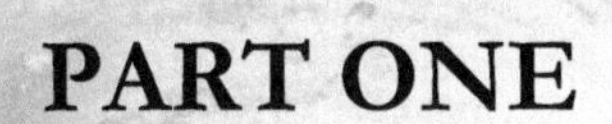

PART ONE

Safety Leadership Concepts

The most influential people strive for genuine commitment; they don't rely on compliance techniques that only secure short-term persuasion.

- Mark Goulston

1

COMPLIANCE, COMMITMENT & LEADERSHIP

Why do you do what you do? We all have a reason or purpose for taking any action, no matter how significant or trivial that action may be. We view the world in a certain way, gain experiences throughout our lives, and develop a set of values and belief systems. Ultimately, our actions are mostly driven by our beliefs.

In his seminal work, "Start With Why", Simon Sinek explores this concept in great detail.[1] He introduces the notion of the Golden Circle and makes the case that the most inspirational leaders don't focus on WHAT they do or even HOW they do it, but WHY they exist...their central purpose or beliefs.

It is my contention that when it comes to safety performance, most successful organizations have a core group of leaders at every level who understand and embrace this concept. Specifically, the leaders within these companies always start with a WHY of caring.

Caring. It's a simple word. In my experience, it is the single most important attribute of leadership needed to develop a healthy safety culture. Sadly, it is frequently absent in many

---

1 *Start With Why. How great leaders inspire everyone to take action.* p39. Simon Sinek. Portfolio/Penguin. London. 2009.

organizations. It is almost as if personal expressions like empathy and compassion do not belong in the workplace.

Organizations that seek to achieve true safety excellence realize this level of performance can only be attained with strong leadership. However, not everyone would describe strong, effective leadership in the same way.

When it comes to safety, there are two prevalent leadership philosophies. There is a stark contrast between the resulting safety cultures. We can better understand the differences by realizing each is grounded in very different motives (personal WHY's). One approach emphasizes control, while the other starts with caring.

Control = Compliance

Some managers define "strong leadership" as carrying a big stick. These managers believe any time there is an injury or near miss, their principal responsibility is to hold people accountable. In practice, this means the primary reason they have any safety conversation is to exert more control.

These managers believe if people would simply comply with the policies, rules, and procedures, then no one would get hurt. Armed with this reasoning, they strive for greater **control** by criticizing actions inconsistent with established policies. Safety conversations center on correcting errant behaviors through counseling or discipline.

This safety approach leads to a *Culture of Compliance.*

The graphic on the next page demonstrates the actions leading to this kind of safety culture. It is summarized in the following statement:

If the reason (WHY) you have any safety conversation is to exert **control,** the approach will be to criticize (HOW) and seek compliance through correction (WHAT).

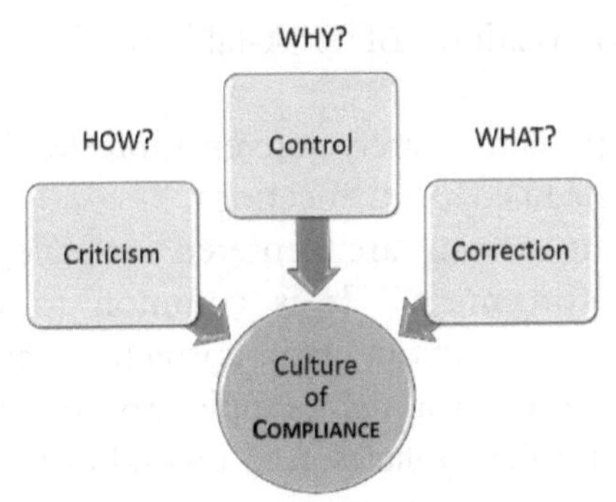

A Culture of Compliance results in a *false sense* of improved safety performance because many incidents are driven underground. The official safety numbers may look good. However, the number of unreported near misses and unrecorded minor injuries are indicative of an insidious safety culture. Because the causes are never acknowledged and addressed, they accumulate until a significant event occurs.

Let's take a deeper dive into the attributes of this culture. Take a look at the figure below.

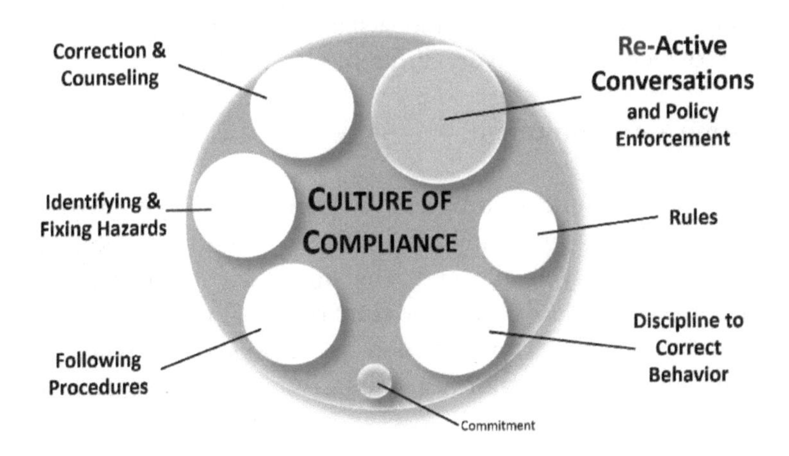

The predominant feature of organizations with this culture is most safety conversations are **reactive.** This simply means there is frequently a triggering event (an injury, a reported near miss, an observation of risk-taking) for most safety conversations.

Almost all proactive safety conversations tend to be limited to regularly scheduled safety meetings.

Management actions are centered on rule, policy, and procedures enforcement. It is common to react to any perceived errant behaviors with correction, counseling, or discipline. In this environment, employees are told they *have to* comply with all rules, policies, and procedures – if not, there are immediate and negative consequences.

Consistent with managing for safety, the lion's share of the organization's time and resources are spent on identifying and fixing hazards.

In this kind of work environment, employee commitment is minimal to non-existent. Think about it. If you worked in this environment (where most of the feedback you get is negative and any mistakes are punished), how committed would you be? This form of safety management has been tried over and over with limited results. Many supervisors and managers use this approach because they don't know there is a different (more effective and sustainable) way.

Now let's look at a very different safety culture.

Caring = Commitment

In a safety context, progressive managers define "strong leadership" as providing clear expectations, encouraging constant improvement, and enabling their employees to succeed. These leaders believe their principal responsibility is to understand the error traps and risk-taking factors that may contribute to an injury or near miss. (We will discuss these in more detail later).

The primary reason these leaders have any safety conversation is because they genuinely **care** about their co-workers. They realize process and organizational flaws will only be surfaced and eliminated if there is an environment of trust and mutual learning.

These leaders believe if everyone works together to resolve these issues, tragic events can be avoided. Because they have a caring attitude, these leaders tend to coach when at-risk behaviors are observed. They seek to understand risk-taking factors. Safety conversations are focused on collaborating to resolve issues that could lead to an injury.

This philosophy of safety fosters a *Culture of Commitment.*

The illustration below shows how this kind of safety culture is formed. It can be summarized by the following statement:

If the reason (WHY) you have any safety conversation is because you **care**, the approach will be to coach (HOW) and seek commitment through collaboration (WHAT).

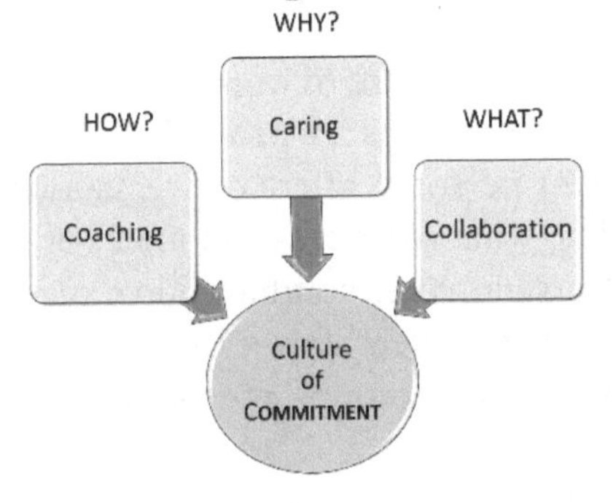

A Culture of Commitment is the foundation for achieving world-class safety performance. Near misses and any injuries, regardless of severity, are reported and thoroughly investigated. Importantly, frequent pro-active safety conversations are integrated into every leader's regimen.

This is also a culture of learning and improvement. By constantly taking care of the small things, the big events are a rare occurrence.

The figure below describes the main characteristics of this culture.

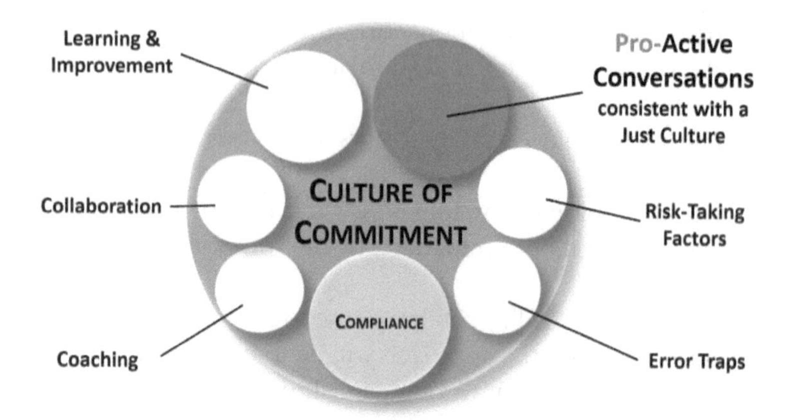

The predominant feature of organizations with this culture is most safety conversations are **proactive**. This means leaders facilitate frequent personal safety conversations with the goal of understanding what *might* cause an injury or an event.

These conversations are conducted in a manner that builds and promotes trust, which allows for a dialogue where risk-taking factors and error traps are identified and solutions implemented.

Leadership actions are centered on coaching for inaccurate perceptions and risk-taking habits. Leaders also seek to collaborate with employees to make it easier to perform a task safely. Because these organizations are leading for safety, a principal objective is for everyone to constantly learn and improve.

Note that compliance absolutely exists in a culture of commitment. After all, we need people to be in compliance with the safety rules.

However, in these organizations people are compliant not because they are told they have to, but because they *want* to! Why? They simply recognize it is the right thing to do!

The only sustainable path forward

Now that we understand the principle differences between these cultures, let's see the safety performance we can expect with each approach.

The figure below displays reported safety incidents over time for both of the cultures discussed earlier. It supports this hypothesis:

If you have poor safety performance, you can almost certainly gain some immediate improvement simply by making sure policies and safety rules are clear and everyone follows them. Your strategy is to manage for compliance.

However, notice what happens after the initial gains. Further improvements are incremental and unsustainable. In addition, incidents don't necessarily decrease, *they just don't get reported.* Ironically, when some organizations get to this point, they try to drive improved safety by doing more of the same. Perform more audits. Increase discipline. Write and enforce more policies. But it doesn't work.

The only way you can create an injury-free workplace is by leading in a way that fosters employee engagement and commitment.

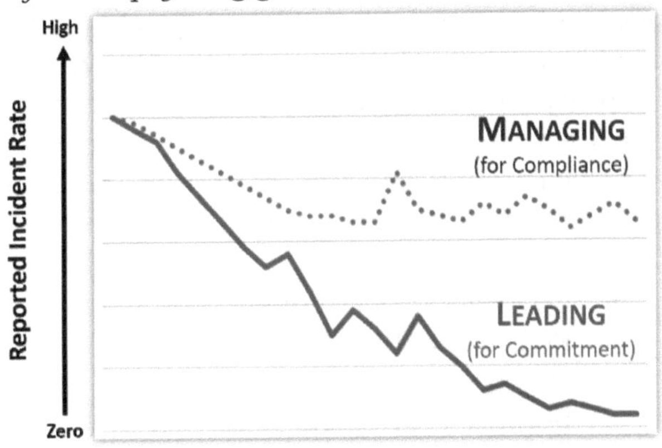

You may be skeptical about this argument. To prove my point, let's perform a "thought experiment".

Imagine I give you a magic wand. This magic wand does only one thing. But it is extremely powerful. If you wave this wand over your workplace, magically the following happens:

100% of the employees will follow 100% of all safety rules, policies, and procedures 100% of the time. No exceptions.

Here's my question, "Can anyone still be hurt?"

When I conduct this exercise with clients, the unanimous answer is "Yes, of course."

Everyone recognizes compliance alone (while important) is not the only factor that determines whether employees will get injured. Other significant factors include mindset, situational awareness, and the willingness to speak up to one another – to name a few. And many of these factors can be nurtured through leadership and the right conversations. I will discuss in more detail in later chapters.

Safety Leadership Continuum™

Very few safety cultures can be characterized as either entirely compliance or commitment. Most organizations lie somewhere on a spectrum.

Leadership is a powerful driver of any safety culture. An assessment that measures employee perceptions of leadership style can provide key indications of where the organization fits on a scale I have named *The Safety Leadership Continuum*™.

I developed a simple survey which measures perceptions on these eight dimensions of safety leadership on a 5-point Likert scale:

- ✓ Personal Conversations
- ✓ Proactive vs Reactive
- ✓ Coaching vs Criticism
- ✓ Improvement Mindset
- ✓ Active Risk Reduction
- ✓ Caring vs Controlling

- ✓ Speaking Up
- ✓ Self-Efficacy

Listed below are four of the eight survey questions:

Personal Conversations

"When was the last time you had a personal conversation (one-on-one) with your manager or supervisor about safety?"

1. More than a month ago
2. One month ago
3. Several weeks ago
4. One week ago
5. Within the last few days

Coaching vs Criticism

"When it comes to safety, to what extent is your supervisor or manager a coach rather than a cop?"

1. Never a coach. Always a cop. Safety conversations focus almost exclusively on catching people doing things wrong.
2. Rarely a coach. Mostly a cop. Safety is mostly about making sure you follow the rules.
3. Sometimes a coach. Sometimes a cop. I get some positive feedback, but an equal amount of criticism.
4. Mostly a coach. Occasionally a cop. We have some good proactive safety discussions, but we are also expected to follow critical safety rules.
5. Almost always a coach. He genuinely seeks to understand the risk-taking, and we work together to minimize risk. He also holds people accountable if they do not accept coaching.

Caring vs Controlling

"How often does your supervisor or manager demonstrate he genuinely cares about your personal safety?"

1. I don't believe he/she cares about my safety.
2. Rarely
3. Sometimes
4. Most of the time
5. I know he/she cares about my personal safety. This is frequently demonstrated to me.

Speaking Up

"How comfortable are you in stopping and talking to a co-worker if you see them taking an unnecessary risk, even if they are more senior than you?"

1. Definitely uncomfortable. It's not my job to question someone else's work, especially if they are very experienced.
2. A little bit uncomfortable. I would approach a few people I know well, but most of the time I would not say anything.
3. Somewhat comfortable. If someone in my work group was taking a risk that could result in a serious injury, I would speak up.
4. Comfortable. I would say something to any one of my co-workers if they were placing themselves at risk.
5. Very comfortable. I have a responsibility to speak up to anyone – regardless of their authority. I would expect them to do the same for me.

By collecting and analyzing employee responses on each of the eight dimensions, a composite Safety Leadership (SL) Score can be used to place an organization on the *Safety Leadership Continuum*™.

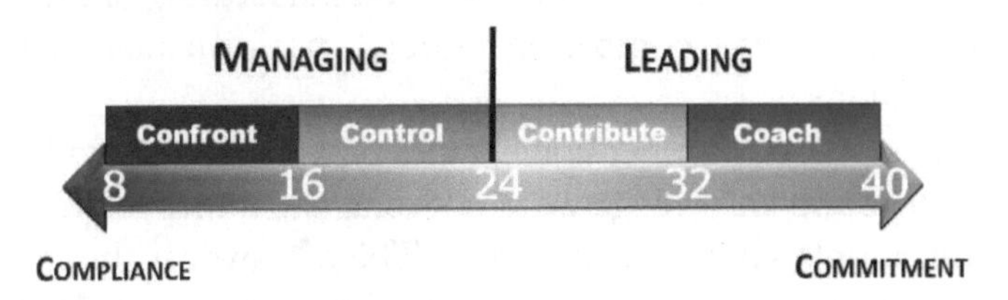

If the SL Score is less than 24, most of your supervisors and managers are using confrontation or control to achieve compliance. They are **managing** for safety.

If the SL Score is over 24, a majority of your supervisors and managers seek contributions from others. They serve as coaches and collaborate on solutions to gain commitment. They are **leading** for safety.

So where do most organizations reside on this scale? At the time of this writing, over 5500 hourly employees from manufacturing organizations have completed this survey. The distribution of the SL Scores from all the surveys is presented below.

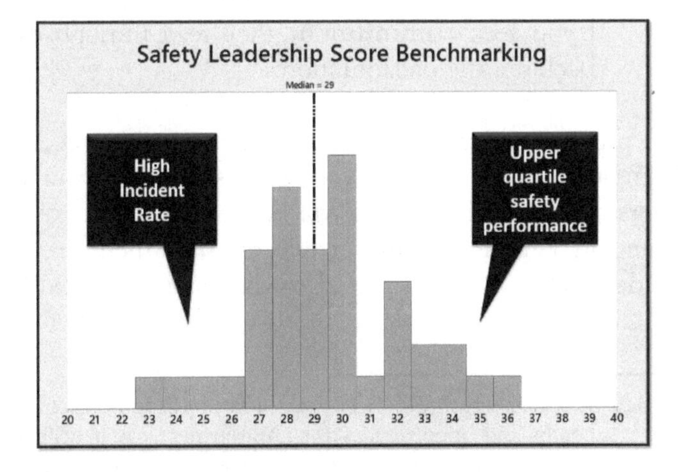

Here is the key finding from all these surveys. **There is a strong relationship between SL Score and safety performance.** The higher the SL Score, the lower the incident rate or injury rate. This observation has been validated within an organization that has numerous manufacturing sites; it is also true when comparing safety performance across organizations.

This is extremely encouraging and powerful. It provides us with profound knowledge we can leverage to improve our own organization's safety performance. The takeaway is this:

The leadership dimensions on this survey are <u>leading indicators</u> of safety performance. If you want to improve safety beyond that which is achievable solely through compliance, develop a plan to improve your employees' experiences on these eight dimensions. In other words, provide the skills for your supervisors and managers to be strong, effective safety leaders.[2]

Everything starts with your WHY. The path to excellence requires leaders to understand their crucial role in determining the safety culture. Leadership style sets the tone for the behaviors that follow.

- If you want **compliance**, then manage through control, criticism, and correction.
- If you seek **commitment**, then lead through caring, coaching, and collaboration.

A principle objective of this book is to provide a collection of simple, practical tools you can use to strengthen your safety leadership skills. These tools will be reviewed and explained starting in Part Two. Before we do this, there are some foundational concepts we need to discuss in the next three chapters.

---

[2] A statistical analysis of the individual survey questions often reveals relative strengths and opportunities for improvement. A benchmarking table for each survey question is presented in the *Appendix* on page 243.

"Would you tell me, please, which way I ought to go from here?"
"That depends a good deal on where you want to get to."
"I don't much care where"
"Then it doesn't matter which way you go."

(Alice talking with the Cheshire Cat)
Alice's Adventures in Wonderland by Lewis Carroll

2

DEVELOPING A SAFETY STRATEGY

What is your safety strategy? When I pose this question to clients, some of them say, "We make sure our employees are trained on all the safety policies, set expectations, then hold them accountable." Some refer to their frequency of audits, behavioral observations, or safety contacts. Others reference the quality and quantity of their mandatory safety meetings, daily toolbox talks, or pre-job briefs. (And some candidly admit they don't have a safety strategy).

None of the actions mentioned above are inappropriate. In fact, if they are simply viewed as a series of things to be done on a master list of safety initiatives, the organization is not likely to achieve its safety goals or reach its full potential.

Most companies spend a considerable amount of time developing, refining, and executing a business strategy. Senior leaders recognize the organization's need to have a sense of direction and purpose that can guide day-to-day activities. Yet many of these same companies do not invest mindshare into developing a safety strategy. These same executives are then perplexed and frustrated when safety performance does not show signs of improvement.

Developing a safety strategy is not difficult. I recommend bringing together "thought leaders" across your organization for this exercise. It also helps to have a structure to follow. In the process I use, this group is initially challenged to establish a safety vision and core principles. Then they drill deeper to find the behaviors that will support this vision. They also consider how to influence key stakeholders. Finally, the process (depicted below) concludes with specific leadership actions.

Let's walk through each of the steps in the process.

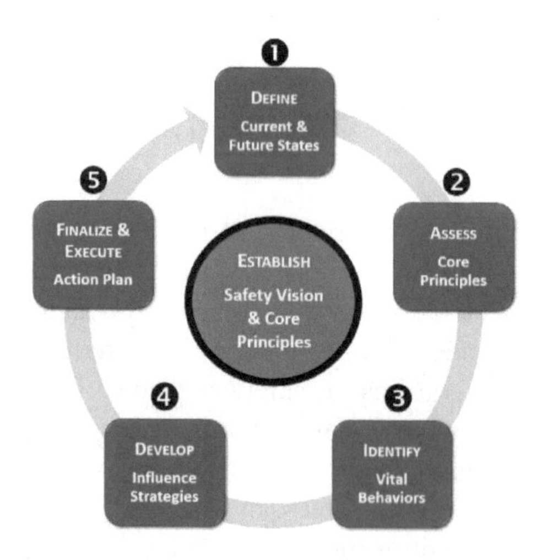

Establish a Safety Vision

A vision statement answers the question, "Where do you want to go?" It explains where your organization is headed. It is considered your North Star.

Some characteristics of a great vision statement include:

- Future-casting (use words like "We will…")
- Clear and Visible (Can you see yourself there?)
- Audacious (Think big!)
- Descriptive (It's only a sentence or two)
- Long-Term (5+ years away)

One exercise that can be used to draft a vision is to break a larger group into small teams. Challenge each team to think about 5+ years from now. Tell them to imagine their company is on the cover of a major industry publication being recognized for their safety performance. Ask them to consider the theoretical headline, what people are saying, what the interviews of employees would reveal, the photos associated with the story, etc.

From this brainstorming exercise, each team is positioned to draft some aspirational statements that might be included as part of the safety vision.

Another exercise that helps to prime the participants into thinking about a safety vision is to read about companies who are acknowledged as leaders in safety for their industry. If those who are in the room have not personally experienced a workplace committed to safety excellence, this is a good way to give them some ideas of what it is like. Once again, this exercise is best completed in small groups. Often I will assign a specific company to each team. I ask them to reach consensus on the key safety strategies or cultural aspects of the company as captured in the article they read, then report their results to the larger group.

All this input is used to facilitate a discussion among the leaders. The objective is to draft a safety vision that is "about 80% right". This vision statement can be completed at a later date with some final wordsmithing.

Here are some safety visions that were developed using this process. (Some of these were edited to keep the organization anonymous).

"We will have an environment of trust and collaboration that enables everyone to go home safe and healthy each day of their working life."

"We believe in working together so no one gets hurt. We use a caring, coaching and collaborative approach to reduce and eliminate risk. We are empowered, engaged and committed to each other's safety and health at work and at home."

"We care about our work families' health and safety. Through a culture of empowerment and engagement, we trust everyone will speak up to eliminate injuries."

"We will always look out for one another because we care. We will collaborate to reduce or eliminate risk. We will all contribute to a safe work environment. We will do these things so we can all return home safely every day."

"We will win together by using our core values to develop and nurture a safety culture built on caring, empowerment, trust and collaboration. Through these values and a strong commitment to the health and safety of our work family, we will achieve safety excellence both at work and at home."

Establish Core Principles

Core principles are shared ideas or beliefs which guide the members of a community in their decisions and actions. They are used to clarify and support the organization's vision.

In order to develop this list, here is the challenge given to the group: "Imagine an ideal world where we all work in an injury-free environment. In order for this to become reality, what **beliefs** or **shared ideas** would have to be held by everyone?" Participants are asked to consider the culture survey questions[3], as well as what they learned from reading about some of America's safest companies. With this framing, it is not unusual to get a dozen or more beliefs or shared ideas that are candidates to be considered as core principles.

Examples of the core principles derived from this exercise are listed on the next page.

(All statements start with, "We believe…")

---

[3] Some of these questions were discussed in Chapter 1, pp 8-10.

- In openly receiving and providing feedback, focusing primarily on positives rather than negatives

- A caring culture goes beyond the workplace to our homes and community

- Safety is never compromised over production. We can replace products or equipment, but not people

- Everyone recognizes hazards or risks and takes responsibility to minimize them

- Everyone has an equal voice in safety and is comfortable in having conversations (even *up*)

- All injuries are preventable

- Everyone has an obligation to speak up and stop work if they see someone taking an unnecessary risk

- Safety concerns are addressed and feedback is provided in a timely manner

- In providing frequent safety coaching and feedback

- Everyone makes mistakes. We collaborate on ways to prevent re-occurrence

- There is a genuine sense of caring and a high level of trust among employees

- Effective leaders conduct frequent proactive safety conversations

Define Current & Future States
Assess Core Principles

The next two steps are closely related and are accomplished at the same time. In essence, the group has already portrayed the future by painting a vision and by establishing core principles which will support this vision.

Participants assess to what extent their organization currently embodies the core principles (of their future state). This is accomplished by using a worksheet like the one shown below.

Individual Core Principle Assessment Worksheet

Core Principle	100%	90	80	70	60	50	40	30	20	10	0%
1			X								
2						X					
3				X							
4								X			
5						X					
6				X							
7					X						
8									X		
9							X				
10						X					
11			X								
12										X	
	ALWAYS	% of time we currently exhibit this principle									NEVER

Now the group needs to reach consensus on the potential impact (High-Med-Low) on the safety culture or safety performance of the organization IF each core principle was exhibited more frequently or more consistently by employees. This impact assessment is plotted versus the % of time each core principle is currently exhibited.

A typical work product is shown on the next page.

The obvious opportunity space is in the upper right quadrant. These core principles are not currently exhibited very frequently, yet are determined to have a high impact on the safety culture. I refer to these as **Target Core Principles**. These are the ones the group will carry forward as they continue to develop their safety strategy.

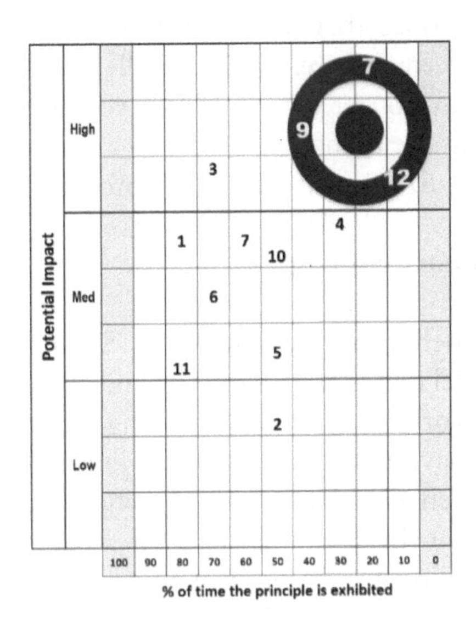

Identify Vital Behaviors

Vital behavior is a term used by the founders of VitalSmarts™. They define a vital behavior as a high-leverage action that (if routinely enacted) will lead to the results you want.[4]

The group is introduced to the concept of a *positive deviant.* These are people who are successful at doing something where others are not. The question is, "What specifically do they do to be successful?"

The group is divided into small teams for this exercise. Each team is assigned one of the target core principles. They

[4] *Influencer. The Power to Change Anything.* Patterson, Grenny, Maxfield, McMillan, & Switzler. McGraw-Hill. 2008. pp. 23-41

are asked to think of a person who exemplifies this principle. By their actions, they are a role model - a positive deviant.

The teams brainstorm specific vital behaviors the positive deviant exhibits which reinforce the target core principle. These behaviors enable them to be much more successful than their peers.

The lists of vital behaviors are combined, and consensus is reached on the ones that have the potential to significantly impact the safety culture. The challenge, of course, is determining how to get others to act consistent with what positive deviants are doing.

This takes us to the next step in the process.

Develop Influence Strategies

B.J. Fogg, a behavior scientist from Stanford, created a behavioral model that is easy to understand and apply. We can use this framework to assist in developing influence strategies for the vital behaviors we identified previously.

The Fogg Behavior Model (FBM)[5] shows three elements must converge at the same moment for a behavior to occur: Motivation, Ability, and Trigger. When a behavior does not occur, at least one of those three elements is missing. Fogg simplifies this concept into the following formula:

where: **B** = behavior
m = motivation
a = ability
t = trigger

---

[5] *http://captology.stanford.edu/projects/behavior-wizard-2.html.*

If we want a vital behavior to occur, we need to assess whether any of the elements are missing and develop plans to fulfill these needs. Do we need to increase the number or effectiveness of the **triggers**, enhance a person's **ability** to do the task, or amplify their **motivation**?

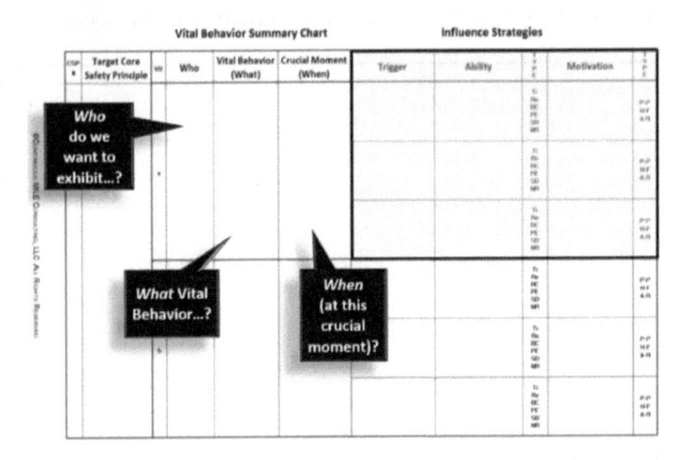

The participants go through a series of brainstorming sessions where each vital behavior is assessed using the FBM. Small teams work on specific behaviors of both workers and supervisors to come up with a plan to influence others to exhibit the behaviors of the positive deviants.

Finalize & Execute the Action Plan

The last step is to convert these influence strategies into an **action plan** with clear assignments and timing.

In addition to the strategy session work, there is another source of potential leadership actions: key findings from the safety culture survey described in Chapter 1. Be sure to fold these into your final action plan. In my experience, the areas of opportunity most often identified in the safety culture survey are:

- Supervisors conduct proactive safety conversations infrequently

- A significant number of employees are not comfortable speaking up to their peers

This step is where many organizations fall short. Even if they spend the time to develop a safety strategy, it means very little if leadership fails to execute the action plan *(see figure below).*

Safety Leadership Action Plan

CSP# 1 — Safety is never compromised over production.
We can replace products, but not people.

	SUPPORTING BEHAVIOR WHO WHAT WHEN	Action Ownership	Comments	Timing
ABILITY				
1				
2				
MOTIVATION				
1				
2				

The payback for all your efforts should be clear. If you have been struggling to significantly improve safety beyond what can be achieved through compliance, you now have a plan which should give you confidence you are working on things that will make a difference. It is an action plan based on two highly credible internal sources:

1. The voices of your employees obtained from the safety culture survey
2. Actions directly linked to your safety vision and target core principles

Now that you have a strategy and an action plan which addresses the most urgent gaps, it is time to turn our attention to some fundamental safety leadership concepts. **Unless you understand and embrace these behavioral and leadership principles, you will not be effective in improving your safety culture.**

"Experience is simply the name we give our mistakes"

\- Oscar Wilde

3

MISTAKE MAKERS & RISK TAKERS

Are you good at multi-tasking? When I ask this question to participants in my workshops, I almost always get one or two people who will nod or raise their hand. In reality, it is extremely rare for someone to be able to truly multi-task (defined as carrying on two conscious activities at the same time). One study places this number at about 2% of the population. Scientists have discovered the reason these exceptional few people can multi-task is their brains are wired differently.

A simple exercise developed by Dave Crenshaw[6] demonstrates your inability to multi-task. Participants are given a sheet with the words, "Multitasking is worse than a lie". In the first pass, each person is timed as they re-write this sentence, followed immediately by writing the numbers 1 through 27. In the second pass, individuals are asked to switch between writing the letters and the numbers, completing both strings of characters and numbers in this way. The time to complete this task is also recorded. A typical result: it takes about *twice as long* to complete the assignment when alternating between letters and numbers.

---

[6] *https://davecrenshaw.com/downloads/multitasking-exercise-v2.pdf*

The implications for workplace safety are obvious. Would you want to work with someone who is driving a heavy piece of mobile equipment while trying to talk on their cell phone? Their reaction time could be critical if a co-worker unexpectedly stepped in front of their vehicle. Would they be able to stop in time?

Error Traps

Multi-tasking (or more accurately, the attempt to multi-task) is just one of numerous situations or circumstances that increase our likelihood for making a mistake. These situations are sometimes referred to as "error traps".

Listed below are ten prevalent error traps.[7] [8]

- time pressure
- distraction
- frustration
- multi-tasking
- complacency
- vague guidance
- peer pressure
- non-normal
- environment
- fatigue

Error traps play a significant role in the workplace. While many people experience the effects these situations or

---

[7] This is not an "ordered" list. Any of these situations can increase the chance for human error. SafeStart™ uses a condensed list and refers to these as "states". The four states used by that organization are rushing, frustration, complacency, and fatigue.

[8] Tim Autrey discusses how to use a set of Error Elimination Tools™ to mitigate the effects of these error traps in his book: *6-Hour Safety Culture: How to Sustainably Reduce Human Error and Risk, (and do what training alone can't do)*. Human Performance Association. 2015. pp. 250-264.

circumstances can have on their work performance, most mistakes are not life-threatening. However, mistakes in an industrial environment can have catastrophic or tragic consequences.

Supervisors or managers managing for compliance often see mistakes as lapses in attention or carelessness. Their reaction to these errors is to counsel or discipline the individual. They believe the employee just needs to pay more attention and harsh words or disciplinary action will "sharpen their focus". By doing so, the supervisor initiates an insidious set of events sometimes referred to as the *Blame Cycle.*

As the figure illustrates, **counseling or disciplining** an employee who made a mistake reduces the level of trust between the individual and his supervisor. Because the employee's trust level is lower, the employee is less likely to talk with the supervisor. This means the supervisor is less aware of any latent process issues. Error traps within the process are unresolved. Any error traps that are a part of the process accumulate over time.

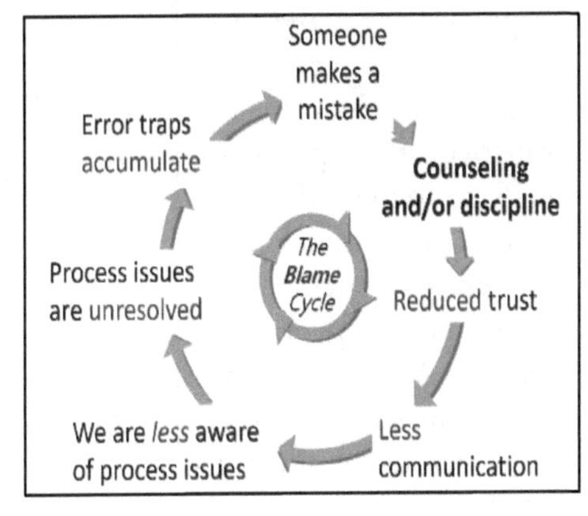

In other words, nothing changes, nothing is identified as a weakness, and there are no improvements. The cycle comes full circle when the next mistake is made (perhaps even by a different person) because the original process weaknesses that contributed to the error were never resolved. Meanwhile, new process flaws may have surfaced.

Compared to managers who are trying to drive safety primarily through compliance, leaders who are fostering a culture of commitment take a very different course of action when mistakes occur. These leaders recognize we are all fallible (error-prone) and inadvertent or unintended actions occur every day.

In this environment, the initial response when someone makes a mistake is to **console** the individual.[9] Leaders espousing a culture of commitment also seek to understand what may have contributed to the error. The primary objective of this discussion is to learn from the mistake and to make improvements to the process that minimize the likelihood of the same error happening again. By taking this approach, these leaders are creating and sustaining a *Trust Cycle.*

Note that an initial response of *empathic understanding*, coupled with a shared desire to improve the process, leads to increased trust, more communication, greater awareness, and collaborative problem-solving. As a result, the same mistake is <u>less likely</u> to occur in the future.

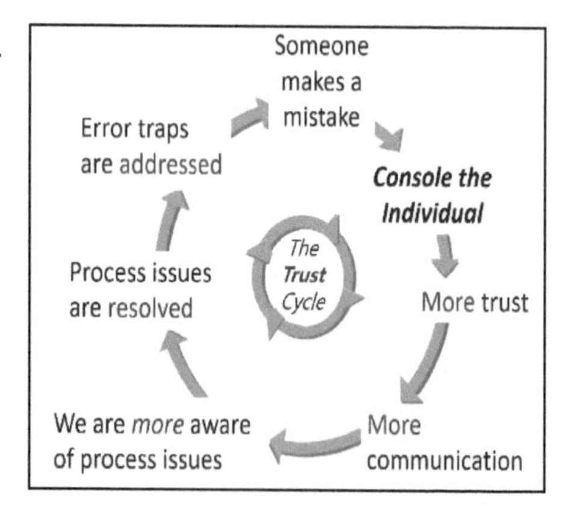

Let's examine a situation (based on actual events) where an operator made a mistake. Because the supervisor responded appropriately, the outcome was favorable from every vantage point.

A manufacturing process was designed so two large machines shared a raw material preparation area. Unfortunately, about every six or eight weeks a costly mistake was made. An operator in the prep area would send a vessel of mixed raw materials to the wrong machine for the next processing step. By the time the error was recognized, a large

---

[9] Consoling is preferred when addressing true human error. Counseling or discipline is appropriate for reckless or certain repeat behaviors. Consoling, coaching, and discipline are discussed in Chapter 4: *What is a Just Culture?*

volume of product would have to be discarded. When a team looked at the historical data, they noted this error was made by all operators at about an equal rate.

Perplexed, a supervisor asked one of the operators to show him step-by-step how he selected the raw materials, mixed them, and sent them to their intended destination using a distributed control system.

The "aha" moment came when the 59 year-old operator demonstrated how he used a mouse to click on the machine system where the raw materials were needed. The computer screen had one line for each machine. The actual size of the lines looked something like this:

Machine 1

Machine 2

The cause of the error (and the solution) were immediately obvious. Imagine if you were an operator who was tasked with monitoring the process and processing hundreds of batches of mixed materials each week. As you peer through your reading glasses at the computer screen full of process information and move your mouse over the choices of where to send the materials, what is the likelihood you would inadvertently select the wrong machine? Actually, it's amazing this error only happened every six weeks!

Upon seeing this process flaw, the supervisor *consoled* the operator and solicited his ideas for a potential solution. The main improvement was a re-design of the computer screen to enlarge and color code the font (which only cost some programming time)[10].

In this case, the supervisor entered the Trust Cycle. What followed was more communication, a flow of ideas, a process improvement, and a reduction in the occurrence of this

---

[10] This is a simple example of mistake-proofing or *poka-yoke.* More information on this topic is found in Notes & References, pp. 245-246.

specific error. (The facility went several years without this mistake after these design changes were implemented).

Risk-Taking Factors

Have you ever been in this situation? You hear about someone who got injured. Perhaps it was a co-worker. Maybe it was a neighbor. Or you may have read about the circumstances in the media. Upon learning about what happened, sometimes you say to yourself, "Why would anyone think it was okay to do **that?**"

It is easy to look at someone's decision to take a risk *after* they are injured and question their judgment. Indeed, sometimes people are just plain foolish in their actions. These are the situations where the person was oblivious to their surroundings or ignored all the warning signs that screamed, "This is a BAD idea!" Or perhaps they were even reckless.[11]

It is **not** reckless or foolish behavior that leads to most injuries. It is a combination of subtle forces that contribute to risk-taking.

Here is a story all too familiar to EMT's and first responders.

Greg was running late.

He glanced at the clock on the dashboard of his truck. His son, Travis, was the starting pitcher today for his high school team. Travis had traveled with the rest of his team to the field and was currently completing the last of his warmup pitches. Greg's wife was going to meet him at the game which was scheduled to start in five minutes. Yet he was still fifteen minutes away from the ball field.

As Greg's truck crested a small rise on the two-lane road, he spotted a tractor ahead pulling a cultivator. Greg braked hard and slowed to 20 mph, slamming his hands on the steering wheel in frustration. Another hill loomed in the distance. Greg eased the

---

[11] Reckless behavior is defined on page 62.

truck into the other lane numerous times as he looked for an opportunity to pass the farm vehicle blocking his view. Each time, he retreated behind the tractor as a car approached from the opposite direction and passed by. Finally, Greg saw an opening. Ignoring the double-yellow line, he steered his truck around the lumbering farm equipment and quickly accelerated.

Greg didn't see the oncoming vehicle. It was obscured by a small rise in the road. The last thing Greg remembered on that fateful day was swerving to the left. He watched in what seemed like slow-motion as fence posts were clipped by his front bumper, each one splintering like a matchstick before disappearing over the roof of the truck. At the end of the fence line stood a large locust tree…

Meanwhile, Travis had pitched several innings and was doing well. That's why Travis was surprised when his coach came to the mound in the middle of the third inning. The coach asked for the ball and told Travis to go see his mother, who was sitting in the stands behind the dugout. She was wiping tears from her cheeks while pressing a cell phone to her ear…

Do you see the error traps of time pressure and frustration in this story? Both of these certainly played a role in the outcome. But something else contributed to Greg's fateful decision. In the moment before he crossed the center line, Greg believed he could easily overtake the farm vehicle. He was confident there wasn't much risk in passing the slow tractor on a double yellow line. Unfortunately, he was wrong. Dead wrong.

Risk-Taking Factors are situations or conditions which may cause someone taking an unnecessary risk. Four prevalent factors are listed below.

- Perception vs Reality
- Repeated Success
- Extra Effort Required
- Resources Unavailable

To better understand these factors, we will provide a definition for each situation or condition and provide some examples.

Perception vs Reality

This risk-taking factor is present when the *perceived* risk is lower than the actual risk of a task. There is a mismatch between the person's view of the amount of risk (low) and the real risk (higher).

In this instance, people take risks because they have an inaccurate perception of the real risk. They convince themselves, "I'm not likely to get hurt doing it this way." They don't think anything will happen to them. They recognize there is some risk, but underestimate this risk and/or overestimate their ability to overcome anything which may get them hurt. Their judgment on the amount of risk is flawed.

How did Greg's perceived view of risk impact his decision? He was driving a lighter, faster vehicle compared to the slow-moving farm vehicle. He had (successfully) overtaken many other vehicles on a two lane road before. He had driven this same road many times before and knew it was lightly traveled. Combine these perceptions with his self-imposed time pressure and the frustration of being stuck behind a tractor. Greg weighed all of these factors and made the fateful choice to take the risk of passing on a double yellow line.

I'd like to share a personal example of how inaccurate *perceived* risk significantly contributed to a poor safety decision.

Recently I contracted with a tree service to remove some dead ash trees on my property. I researched options and decided on a service that promoted their experience and safety record. Early one weekday morning, three young men (who looked to be in their late twenties) arrived to do the work. I was especially keen on how they would be felling and removing the trees safely. I had a brief conversation about the job with the team leader and clarified which trees were to be taken down. Then I stood back to observe.

Initially, I was impressed. Each man was wearing steel-toed shoes and long pants with chaps. They proceeded to put on other personal protective equipment. These items included hard hats, safety glasses, and long-sleeved gloves. After the team leader finished fueling the chain saw and filling the oil reservoir, he placed the saw on the ground and reached for the pull cord.

I immediately stepped in and yelled, "Hey, wait a minute! What about your hearing protection?"

I was astonished to hear him casually reply, "Look, we are only going to be here for about 5 or 6 hours."

"Well, you aren't running a chain saw on my property without wearing hearing protection," I said.

The team leader rolled his eyes as he looked up at me from the chain saw sitting on the ground in front of him. "We don't have any hearing protection. So do you want this job done or not?" He was clearly upset.

"Wait here," I ordered. "I'll be right back."

A few minutes later I returned with two sets of ear muffs and a pair of ear plugs.

"Here you go, guys. I really care about your hearing. Here's the deal. You can either wear these or come back another day."

They reluctantly accepted and used the hearing protection and safely completed the job.

What was the <u>perception</u> of these young men about their hearing? Of course – they would always have the same hearing when they are sixty years old. Even if they are around chain saws and wood chippers for many years!

Here is what these men didn't know about the guy who insisted they wear hearing protection. I am profoundly deaf in one ear. I have been one-sided deaf since I was six years old. (I lost my hearing because I had a severe case of the mumps).

I never saw my deafness as a handicap. I simply adapted and learned to position myself so I was in the best possible location to hear most of what was being said. This wasn't always possible, of course. There were quite a few times when people on my "deaf" side approached me or asked me a question, and I just didn't respond.

I'm sharing the story about my encounter with the tree service crew and my deafness for two reasons.

First, it underscores how all perceptions are personal. The perceptions of those young men about the risk to their long-term hearing were wrong – but these perceptions were theirs. They weren't bad or careless people. After all, they knew enough about safety to wear the other personal protective equipment. They simply needed someone to point out the risk they were taking by not wearing hearing protection. And they needed someone to **care** enough about them to insist they do the right thing. I happened to be the person to give them a different perspective.

Second, my life-long experience with being one-sided deaf has impacted me in ways I am just now beginning to appreciate. I am a champion for hearing protection and an advocate for mitigating hearing loss. Whenever I see someone in an environment where they should be wearing hearing protection, I always approach them and point this out and/or offer ear plugs. My hearing loss is undoubtedly one of the motivations for starting my consulting practice. I believe more can be done to prevent not just hearing loss, but any injury that might impact lives or livelihoods.

I'm also teaching my grandson the importance of protecting his hearing whenever he is exposed to any loud noises. After all, it's never too late to change perceptions. And it's never too soon to create experiences that reinforce perceptions!

Repeated Success

This risk-taking factor is present when a risky task is completed many times without incident.

If we do something repetitively over a period of time and the results are what we expect, these behaviors are likely to become habitual. We don't even have to think about doing a certain task because of the muscle memory that has developed. Our actions become almost mechanical – we don't give them much thought.

You probably have many personal habits. Examples could be brushing your teeth, making coffee, tying your shoes, or taking medication at a certain time of day. You do these things every day. They are a part of your routine. And you don't have to spend much brain power completing these tasks.

One way to discern whether the set of behaviors has truly become a habit is to ask yourself, "Can I do this and think about something else at the same time?" (Recall the definition of multi-tasking: carrying on two *conscious* activities at the same time). Your set of behaviors is considered a habit if you can consciously perform another activity at the same time.

Some repetitive behaviors in the workplace are good. Any set of repeating actions that enable us to be productive while keeping us and others out of harm's way are safe work habits. These should be encouraged and reinforced. Safe work habits are a key strategy for mitigating error traps, which were discussed earlier.

Examples of safe work habits include: using three points of contact when ascending or descending, being alert for line-of-fire situations, moving your eyes before you physically move, buckling up when you are in any moving vehicle, making eye contact with the other guy, looking at where you are placing your hands, and many others.

But there are also repetitive behaviors where each time we do them, we are taking a risk. We refer to these as risk-taking habits. Many times people who have developed risk-taking habits are either unaware of the actual risk or they don't believe the risk they are taking is significant.

Here are a few examples of risk-taking habits: texting or talking on a cell phone while walking (or driving!), taking stair steps two-at-a-time, crossing a street without looking both ways, leaning back on a chair so it is only on two supporting legs, using a chair with casters as a step stool, backing up a vehicle without first looking, and so many more!

There is a strong connection between perceived risk and habits. In fact, almost all risk-taking habits initially develop from our perceptions. If I perceive there isn't much risk in taking a certain action, then proceed with doing a task that way and nothing bad happens, my belief is reinforced. This encourages me to perform this same task again in a way that places me at risk. With each successive (and successful) completion of this task, I receive confirmation there isn't a significant amount of risk involved in doing it that way. After all, I haven't gotten hurt (yet!).

The bottom line: unless and until I have an experience strong enough to change the perceived risk, my risk-taking habit will likely continue.

The following story, based on actual events, illustrates how an inaccurate perception and a risk-taking habit combined to create a near-miss.

Luke had 14 years of experience in the industry and had recently been promoted to supervisor. He came out of a meeting and walked onto the operating floor. Before he could say anything or respond, Luke saw crew member Blaine reaching into a rope cluster near a paper roll to remove some scraps of paper.

This paper roll runs at a very high speed. It is also very close to the ropes. A balancing weight also protrudes from this roll, making the place where Blaine placed his hand a very tight space with a number of fast moving parts.

Luke waited until Blaine withdrew his hand from the rope cluster before shouting to him and motioning for him to come over. Luke realized if Blaine was distracted while he was in this position, he might inadvertently move his hand into the moving paper roll.

Blaine had been working on the machine for ten years. He was a hard worker who was quick to respond to a machine break. Blaine helps his father on a farm and hoped to take up farming full time in the future.

Luke recalled just last month another employee received a laceration when he reached into a slitter blade assembly on another machine. He approached Blaine to talk about the near miss he had just witnessed. Luke really wanted to understand why Blaine had taken this risk.

Let's start by looking at the situation from Blaine's perspective. Here's what we know from reading about this event. Blaine is a hard worker who wants to do a good job. He is an experienced operator. He has a farming background and therefore is accustomed to being around machinery.

But there are some important things we don't know. There are other factors that almost certainly contributed to this near miss. But unless Luke has the *right kind of conversation* with Blaine, he is not likely to discover what these are. (We will learn about these other factors later).

The early exchange between Luke and Blaine might have gone something like this:

"Hey, Blaine. Do you know why I asked you to stop and come over to talk?" Luke tried to maintain his composure, despite being upset at what he just saw.

"Not really," Blaine shrugged. "Why?"

"Well, I saw you reach into that tight area by the ropes and pull out a scrap of paper. Frankly, it scared me," Luke said. "You know I really care about you and your safety. Can you help me understand why you reached in there?"

"I don't know, I didn't even think about it. I must have done it a hundred times that way. I've worked around all kinds of machinery - not just here, but on the farm. Besides, I'm really careful…"

Blaine's response gave Luke two big clues about what may have led him to take a risk and place his hand in a tight area. Blaine's perception was there really wasn't much danger in performing the task that way. And over time, his belief was positively reinforced until it translated into a risk-taking habit.

Extra Effort Required

This risk-taking factor is present when completing a task safely requires more time or effort than taking a *perceived* small risk.

In an ideal world, we would design processes that enable workers to complete their assigned tasks efficiently and safely. If you are involved in a greenfield start-up, you may have the opportunity to consider these things in the design engineering phase. The reality for most manufacturing sites is they have evolved over many years – in some cases decades or longer. This means obsolete equipment is removed or replaced, new processes are installed, or perhaps the work flow is altered to accommodate changes in product lines.

All of these modifications and changes inevitably create safety roadblocks for those doing the work. These changes may make it more difficult to reach a valve or turn a wrench. Blind spots could be created where previously there were none. Previous ergonomic designs for repetitive tasks may have been eliminated. And so on.

Another common reason that people take shortcuts is the tools or equipment needed to safely perform a task are not readily available. I facilitate a workshop where participants learn how to conduct an effective safety conversation.[12] Here's a question I ask the workshop attendees:

"How many of you have ever used a screwdriver as a pry bar at home?"

Typically, 80% or more of the participants raise their hand.

"Next question," I say to the group. "And *why* did you use a screwdriver when you knew it was not the right tool for the job?"

"Because," several people quickly offer, "The pry bar was in the __________ and I was in the _________."

"Exactly. You were took a risk because the best tool to do the job safely was not close to where it was needed."

Here is another example of how people are tempted to take a risk when it takes additional time to perform a task. It is based on actual events.

Cliff was working by himself emptying drums that contained spill absorbent material.

On this particular occasion, Cliff noticed the bag was overfilled. He saw it was going to be very hard to pull the bag out of the container. Cliff knew it was going to be a heavy lift, but thought he could handle it by himself. Besides, he thought, there isn't anyone else around to help me. I would have to go all the way to the other end of the building to ask someone for assistance.

Cliff attempted to remove the heavy bag by straddling the drum and pulling it in an upward motion. When he did, the bag did not budge. He immediately felt a "pull" in his back. After taking a few minutes to collect himself, Cliff set the drum on its side. Then he pulled on the bag. By using this technique, he was able to get the bag out of the container.

---

[12] A five-step process for having the **right** safety conversation is described in Chapter 21.

Once he returned to the loading area, Cliff weighed the bag. He confirmed it was indeed exceptionally heavy - around 125 lbs.

Cliff notified his supervisor about what had happened. His back was very sore and he was concerned he would need to take some pain medication at a minimum. It was already starting to stiffen up.

In this case, the risk factor was not the lack of the right tool, but the lack of help…a second person was needed for the task. Rather than swallow his pride and take the time to get help (either personal or mechanical assistance), Cliff soldiered on. You can almost hear him briefly going through the decision-making process in his head:

"It would take too long to…"

"It's only one bag…"

"Other guys have probably lifted a bag this heavy before…"

"…"

Let's revisit the conversation between Luke and Blaine. Recall that Blaine explained to Luke "…besides, I'm really careful…"

"Well, being careful isn't enough, Blaine. All it takes is one time when you misjudge where your hand is placed or don't see everything. And in that moment, you could lose a finger or a hand…or worse. I'm curious why you didn't use the gripper tool specially designed to remove those scraps of paper. By using the gripper tool, you can keep your hands out of those dangerous, tight spaces."

"Well, I'll tell you why," Blaine replied, with a hint of anger in his voice. "We've been trying to get this machine up and running for over an hour. I'm hot and tired and more than a little frustrated. We were close to getting it going again, and I didn't want a scrap of paper to break it down. I just want to sit down for a few minutes and have a cold drink. And look around - I don't see one of those gripper tools anywhere. It seems like they are never where you need one!"

As he finished talking, Blaine pulled a cloth from his back pocket and wiped the sweat from his brow.

Now we have a more complete picture of the risk-taking factors which influenced Blaine's decision. Do you see the situation where extra effort was required to perform the task? The gripper tool was nowhere near its intended point of use. But also notice some of the error traps discussed earlier: *time pressure, fatigue, complacency, frustration, and environment.* In combination with his perceived level of risk and his success in doing this task many times before, Blaine decided to take the risk of placing his hand in a tight space with moving components.

I'm willing to wager if you were in Blaine's shoes (not only in his circumstances, but also with his background and life experiences), you would have been tempted to take the same risk he did at that moment.

My argument: what initially looks from an outsider's perspective as an irrational decision by someone is often better explained by the human conditions we all share. One of these is fallibility (the propensity to make mistakes). The other attribute is the ease with which we are persuaded to take risks (by people, situations, or things).

I will discuss these phenomena throughout this book. It is critical we understand what these forces are and what we can do to mitigate them, especially when it comes to safety.

Resources Unavailable

This risk-taking factor is present when there is no way to safely complete a task without assistance.

Imagine you are walking in a facility and see one of your co-workers, Kenny, near the top of a tall ladder. Kenny is working on the second floor of the building and the ladder is only a few feet from an open wall. This violates the policy for tying off when working above certain heights. You stop and ask him why he isn't wearing a personal fall arrest system and isn't tied off.

Kenny quickly responds, "I know the policy. But where am I supposed to tie off?" He motions around him. "On this cable tray?"

There are other precautions Kenny could take when doing this job. He could have someone hold the ladder for him. Or he could get an aerial lift to compete the job. The reality is that it is extremely difficult for him to comply with this specific safety policy without additional resources – either equipment or a helper.

I heard a tragic story involving fall protection. Rob and Jerry were millwrights on a maintenance crew. They were also best friends. They did a lot of things outside of work and especially loved to go hunting together.

One day they were assigned to change out a piece of equipment about 30 feet above floor level. Both men wore safety harnesses and tied off while they performed the task. They were working at opposite ends of the large machine to coordinate the overhead crane. Suddenly, Rob heard a loud shout. Then there was silence. Rob hurried to the other end of the machine to see Jerry lying on the floor, still in his harness. He died instantly.

An incident investigation revealed there were no certified/engineered anchors in the area where Jerry was working. Jerry had tied off using the D-ring of his safety harness to a section of angle iron, but it had broken off when he lost his balance.

After this tragedy, Rob became an outspoken advocate for proper fall protection. He made it his personal mission to make sure there were anchoring points within reach everywhere on the machine where they were needed to safely perform work. As he explained it, "I never wanted to see anyone else suffer the same fate as my friend Jerry because they were unable to securely tie off."

Some risk-taking factors are not physical in nature. Leaders can encourage employees to take shortcuts by giving conflicting directives. Here is an example.

An older facility had two buildings separated by a two-lane public road. Because the shipping docks were close to the road, tractor trailers had to pull into the parking lot on the opposite side of the road, then back across the road into the docks. Truck drivers could not see any vehicles on the road when they backed up because the buildings created several large blind spots. The policy stated drivers should always use a "spotter" to stop any vehicles on the road and guide the trailer as it was backed into the dock.

Several years ago, a car crashed into the side of a trailer that was midway across the road while in the process of backing into the dock. Investigators quickly determined the truck driver had not used a spotter.

At first glance, the culpability would seem to be squarely on the truck driver because he failed to follow the spotter policy. After more interviews were conducted, additional factors were uncovered.

On the day of the incident, one of the warehouse workers had called off. They were operating with one less person. In addition, the production rate was exceptionally high. The warehouse employees were having trouble moving the finished product away from the wrap line fast enough to prevent it from backing up and shutting down.

The supervisor had given the truck driver involved in the incident orders to move several trailers into the loading dock area. The driver asked for a spotter, but the supervisor told him, "We don't have anyone. Just be careful. Besides, there isn't much traffic on the road at this time of day."

The driver objected. The supervisor told him to either get the trailers backed into the dock or he would be considered as insubordinate and would be disciplined.

There it was. The **supervisor** was the risk-taking factor! Faced with this ultimatum, the driver proceeded without a spotter – and a crash was the outcome.

Error Traps and Risk-Taking Factors Combined

As we have seen, risk-taking factors can determine whether someone takes an unnecessary risk. And we also recognized the ten error traps described earlier can make it more likely for someone to make a mistake. In many cases where someone is injured, both are present.

The graphic below illustrates how these precursors to safety incidents or events are connected. Unfortunately, this scenario plays out on our highways every day.

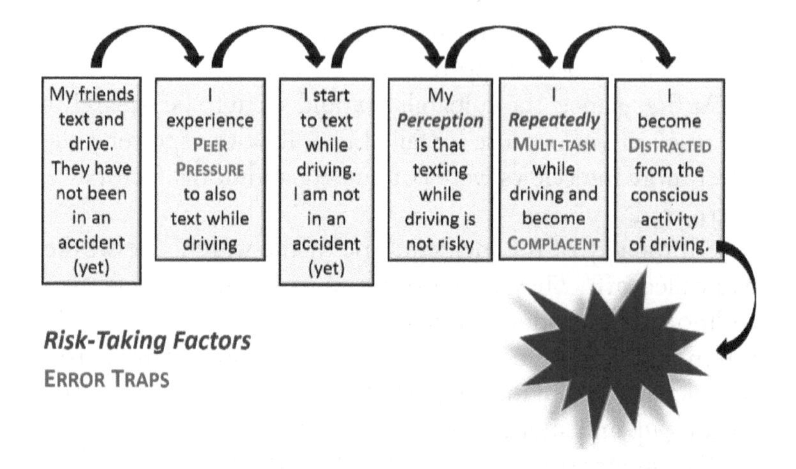

Sometimes a *behavioral pattern* develops as a result of being influenced. We can think of this behavior as a particular kind of habit – one in which the changes in work practices or procedures change gradually over time. That is our next topic.

Procedural Drift

Earlier in my career, I moved to a community in a different state for a new job. While driving to work on the first day, I was involved in some near-miss automobile accidents. I will describe a risky driving behavior which I quickly learned was "the way we drive around here".

Drivers approached an intersection with a traffic light. The green light turned to yellow. As expected, one or two cars entered the intersection while the light was still yellow. But what I observed next surprised me. After the light turned red, the next three drivers continued through the intersection. Remarkably, the cars in the opposing lanes (who had a green light) paused for 3 or 4 seconds for the red light violators to clear, then drove through the intersection. When the light turned red for opposing lanes of traffic, the same behavior repeated. The unspoken norm was that a "red light" meant 3 more cars were allowed to pass through the intersection....the 4th car should stop.

The amazing thing to me was somehow everyone knew this was the rule. At first, I thought this was an isolated incident. As I soon discovered, this happened at every intersection.

Now imagine someone who had never been to this town (me) approaching an intersection - and expecting red means stop and green means go. It took four or five close encounters (of the wrong kind) at intersections with local drivers to figure out what was happening.

I quickly adapted to the local behavior. By the time I arrived back home, I was Driver #3 going through a red light. No consequences. No tickets. In fact, local police cars were following the same protocol! (I learned later three cars was indeed the limit. If the police observed a 4th car driving through a red light, this person was always ticketed).

What was going on here? How could every local person in this large community end up developing and accepting a norm which was clearly violating the standard? One explanation could be the concept of entropy. Here is one definition:

en·tro·py lack of order or predictability; gradual decline into disorder.

Sidney Dekker discusses a specific type of entropy he calls *procedural drift*.[13] He defines it as a mismatch between procedures or rules and actual practice. He claims it almost always exists, and this mismatch can grow over time. Procedural Drift increases the gap between how the system was designed or intended and how it actually works. It tends to be a slow, incremental departure from the designed or intended norm (as depicted in the graphic below).

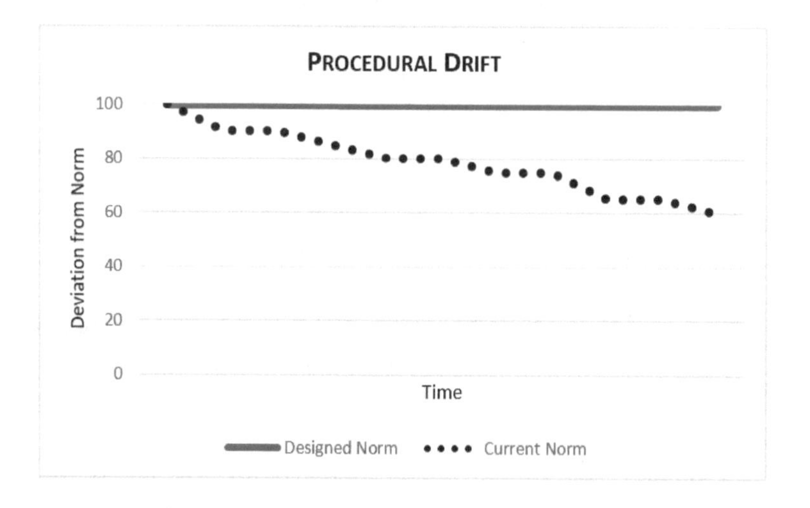

Dekker lists several potential reasons for procedural drift:

- Rules or procedures are over-designed and do not match up with the way work is really done.
- There are conflicting priorities which make it confusing about the most important procedure.
- Past success (in deviating from the norm) is taken as a guarantee for safety. It becomes self-reinforcing.
- Departures from the routine become the *new* routine. Violations become behaviors compliant with local norms.

---

13 *The Field Guide to Understanding Human Error.* Sidney Dekker. Ashgate Publishing. 2006.

From this perspective, the driving practice of "Three-drivers-through-the-red-light" was an example of procedural drift - where departures from the routine were the norm. In this town it was safer to violate the universal "red-light-means-stop" rule. Doing otherwise would increase the chance you would either be rear-ended (if you stopped prematurely on a red light) or broad-sided (if you accelerated too soon from a green light).

In the workplace unchecked procedural drift often results in an undesirable outcome. A defect may be created when work standards, procedures, or rules are not followed. In a safety context, procedural drift could result in a personal injury. In either case, our tendency is to blame people for violating the rules when this happens. During an incident investigation, if we conclude there is a gap between procedures or rules and practice, Dekker challenges us to think about why this occurred.

Dekker asserts it is often compliance (as opposed to deviance) that explains people's behavior: compliance with *norms that evolved over time.* This is quite different than compliance with standard safe operating procedures.

We are also reminded that the continued absence of adverse consequences confirms people's beliefs the behavior is safe. Think about the cars going through the intersection on a red light. If 99% of the time there are no accidents, it must be okay to do this. Further, if law enforcement rarely issues a ticket for this rule violation, it must be acceptable behavior.

Standard Work

Let's consider the implications on establishing and maintaining work standards as a strategy to address procedural drift. One approach would be to write the rules and procedures - and then hold people accountable when they violate these standards. While this may be marginally effective, it is not likely to be sustainable. Why? If we rely solely on people complying with rules, what we get in return is minimal engagement and no desire for improvement.

Standard work can be a good practice - if implemented on the right things and in the right way. Getting those who do the work to agree on what the standard practices will be and co-developing these procedures goes a long way toward making them work. The work standards need certain attributes to be successfully implemented. The procedures should:

- Be as simple as possible
- Be written and explained so they are easy to understand
- Be within the worker's skill and ability
- Include basic mistake-proofing methods

Even if work procedures or standards have these attributes, there is another component needed to make them sustainable and more resistant to procedural drift. We need to recognize because we are asking people to do work differently, there is an element of cultural change. Depending upon the magnitude of change, a detailed implementation plan will need to take this into consideration.

For example, we may need to provide daily coaching and feedback to make sure the procedures stick. Or perhaps establishing a peer-to-peer accountability process would be an effective method to instill the new behaviors.

Who determines what the norms are for any group? Ultimately, it is anyone in an influencing or leadership role. The actions people take are driven by the beliefs they hold. If we want people to take certain actions, we need to provide experiences which reinforce the desired beliefs. We need to think about what we say and do as leaders. For example, do we insist everyone follows a specific safety rule - except when we are behind schedule, and suggest it is okay to take a shortcut "just this one time"? We should not be surprised when someone has a near-miss incident or is injured when they decide to violate this same rule sometime later.

People Problem or Situation Problem?

As a manager, it is likely some of the biggest challenges you face are those you consider to be "people problems." I am referring to the kinds of problems where the character of the individual is *perceived* to be the main reason for a performance issue.

A few examples of *perceived* people problems:

- An employee fails to follow work instructions which results in rework.
- A worker is injured when an unnecessary risk is taken to get the job done.
- A number of employees are perpetually late when submitting expense reports.
- An employee's timeliness in completing some assignments is unacceptable.
- Supervisors do not spend time talking to their employees.

If you were faced with any of the challenges listed, what would you do? Many of us would engage the employee in some form of training, coaching, counseling, and/or expectation setting. In other words, we assume the behavior is largely determined by the individual's character, personality, or mindset. Unfortunately, we frequently overlook the power of "situations" in determining someone's behavior.

A number of years ago, Stanford psychologist Lee Ross conducted a literature review on a large number of studies in psychology. He concluded we have a tendency to ignore the situational forces that shape other people's behavior.

Ross referred to this tendency as the *Fundamental Attribution Error.*[14] We make this error when we attribute people's behavior to **the way they are** (their character) rather than to **their situation** (their environment).

One classic study was set up on a college campus to determine whether people could be persuaded to contribute significantly more to a charitable cause simply by altering their environment.

First, the researchers identified those who were generous (*saints*) from those who were unlikely to give anything at all (*jerks*). They did this by polling everyone in a dorm housing about one hundred students. Each student was asked to assess which of their dorm-mates were most likely and least likely to make a donation to any charity. With this input, researchers classified students into 'charitable' or 'uncharitable' types.

The experiment consisted of making the environment different for two equally split groups. Half of the students received a basic letter announcing the launch of a food drive the following week. These students were asked to bring canned food to a booth at a common student gathering place on campus. The other students received a letter with much more information. Their letter included a map to the exact drop-off location, a specific request for a can of beans, and a suggestion to think about a time when they would normally be passing by the drop-off point, so it would not be inconvenient for them to get there.

These two letters were randomly sent to the saints and the jerks. After the food drive was over, the researchers compiled the names of students who had given food (and those who had not).

Overall, students who received the basic letter were not very charitable. Only 8% of the saints donated. And not a single one of the jerks contributed to the food drive.

---

[14] *The Person and the Situation: Perspectives of Social Psychology.* Lee Ross, Richard E. Nisbett. Pinter & Martin Publishers. 2011.

The students who received the more detailed letter were much more charitable. 42% of the saints donated. But here was the shocking outcome: 25% of the jerks donated food!

The significance of this finding cannot be overstated. The researchers were able to get 25 percent of the *least charitable individuals* in the dorm to donate by **simply making it easier** for them to make a donation. Remember, there was not a huge change to their situation - the letter they received just provided more specific instructions. Their environment (situation) was only slightly modified.

Think about this. In the study, you were 3 times more likely to get the desired outcome by changing *the situation of the jerks* (25% donated) than by relying upon *the character of the saints* (8% donated)!

Many other studies have proven situational forces are powerful when determining a person's behavior.

How can we leverage this knowledge? Don't overlook the person's **ability** to do something. By shaping their environment, we can make it easier to do the right thing and more difficult to do the wrong thing.

Let's do a quick assessment on how we might change the situation to influence the behaviors for each of the examples listed earlier:

- An employee fails to follow work instructions which results in rework. *(Provide clear, easy-to-understand directions with simple diagrams and/or photos).*
- A worker is injured when he takes a risk or makes a mistake to get the job done. *(Address any risk-taking factors. Mitigate or eliminate any error traps).*
- A number of employees are perpetually late when submitting expense reports. *(Make it easy to file a report by minimizing the number of steps, making data entry intuitive, or providing online access).*
- An employee's timeliness in completing some assignments is unacceptable. *(Establish clear priorities, set up self-reporting, offer resources).*

- Supervisors do not spend time talking to their employees. *(Set aside time on their schedules, remove non value-added work, provide a guide for conducting a conversation, track and audit conversations and provide feedback).*

Some human resource professionals advocate employees be evaluated on two dimensions to determine their work performance: will and skill. The first dimension answers the question: "Are they **motivated** to do the job?" The second dimension addresses the question: "Do they have the **ability** to do the job?" And as we have seen, a person's ability can be strongly impacted by their situation.

Let's look at an example where the <u>situation</u> led to a tragic outcome. The following is based on an actual event.

Ed and Phil worked in a paper mill. Ed had over 30 years of experience, having started at an entry-level job right after graduating high school. Now he was a senior operator. Phil hired on a little over a year ago. He had worked construction as well as at another manufacturing plant, but was still learning the papermaking process.

It was 5am when Nick, their supervisor, gathered his crew together for a quick huddle. He gave last-minute instructions to everyone before the machine was shut down for monthly maintenance and repairs.

Nick told Ed and Phil their job was to transport several tote bins of cleaning chemicals to the machine and empty them into a large open tank. He reminded them the maintenance work could not begin until they had completed the cleaning job. After the briefing, everyone scattered to start his or her initial tasks.

"I'll get the chemicals since I've done it many times. I know where these totes are always stored. You get our test equipment and meet me at the tank," Ed told Phil as he climbed on the forklift.

When he arrived in the storage area, Ed noticed it had been re-arranged since the last time he was there to get chemicals. Several overhead lights had burned out which made it more difficult to read the labels on the tote bins. After several minutes of searching, he retrieved two chemical totes and placed them beside the tank.

Phil showed up a minute later. He saw the two totes were not identical, but decided not to say anything since Ed was much more experienced.

"I'll be so glad to get home after this shift," Ed commented to Phil as he cracked the seal and prepared to open the first valve. "I've had to work a double shift the past five days."

Ed emptied the first tote and instructed Phil to place the next one over the tank. Ed opened the valve to the second tote while Phil observed from the forklift.

Within seconds Ed was coughing and wheezing. He fell to his knees and clutched at his chest, gasping for air.

At the same time, Phil instinctively used his shirt to cover his nose and mouth as he crawled off the forklift. He stumbled away from the scene, his eyes watering and stinging, while cursing himself for not remembering where the nearest eye wash station was located. He took a quick glance back at Ed, who was now lying on the floor on his side, coughing violently while attempting to cover his nose and eyes with his hands.

Soon the entire building evacuated as other employees in the area smelled the noxious gas that had enveloped Ed and Phil.

As Ed and Phil were loaded into ambulances, Nick tried to piece together in his mind what had happened. Something had gone terribly wrong.

The incident investigation that followed revealed one tote bin contained a strong caustic while the other contained a strong acid. When these two chemicals were mixed, chlorine gas formed. Ed and Phil were rushed to the emergency room because they inhaled these poisonous fumes.

The initial reaction of many in management was to place blame on these two guys. After all, if they had just been more careful and attentive to the task – and read the labels on the tote bins, this whole nasty incident could have been averted. But guess what? Ed and Phil were fallible human beings (just like all of us). And they made some mistakes (just like all of us). There was a plethora of situations on that fateful day which made it more likely human error would occur.

Here is a partial list of the **situations** or conditions that contributed to this event:

- **Time Pressure** – Ed & Phil were reminded everyone was waiting for them to finish before the maintenance work could begin.
- **Vague Guidance** – Their job was to "transport several totes of cleaning chemicals."
- **Non-Normal** – Ed noticed the storage area had been re-arranged since the last time he was there.
- **Environment** – Several lights had burned out. This made it more difficult to read the labels on the tote bins.
- **Fatigue** – Ed had to work a double shift for the past five days.
- **Peer Pressure**[15] – Phil chose not to say something to Ed about the non-identical tote bins since Ed had more experience.
- **Extra Effort Required** – Ed had to search several minutes in the storage area to find the totes.

As outsiders to this incident, we could simply shrug our shoulders and say, "Oh well. I guess those guys were careless (or inattentive, or lazy, or…). That's why this accident happened." And we would be wrong in our assessment. We would make a Fundamental Attribution Error. By drawing this conclusion, we would be attributing Ed and Phil's behavior to the way they are (their character) rather than to the situation they were in (their environment).

What's the key takeaway? By altering the situation of the job to be done, we have the opportunity to significantly improve an employee's ability to be successful. So before you

[15] Peer pressure (real or perceived) is a powerful reason employees do not speak up to their co-workers if they observe a situation where there is a risk of injury. The topic of "speaking up" will be discussed in Chapter 16.

make a Fundamental Attribution Error (and assume you have a *people problem*), consider you may have a *situation problem.* If it is the latter, take action immediately. Because you almost always have some control over the situation!

There is one more key safety leadership concept we need to review before introducing specific strategies you should have in your **SAFETY LEADER'S TOOLBOX™**. It is the notion of a Just Culture.

"The single greatest impediment to error prevention... is that we punish people for making mistakes"

\- Lucian Leape

4

WHAT IS A JUST CULTURE?

A toddler is left in a car seat on a summer day for several hours and dies after his mother forgets to bring him into the home after returning from a shopping trip. A homeowner falls from an extension ladder and suffers a severe head injury when he loses his balance as he reaches to clean leaves from the gutter. A driver slams into another car at an intersection while traveling over 100 mph killing a teenager.

Beyond the obvious tragedies, in this chapter we will consider the decisions and behaviors which may have contributed to these events in the context of a Just Culture.[16]

Let's start with a definition. A Just Culture is a culture where failure/error is addressed in a manner that promotes learning and improvement while satisfying the need for accountability. An organization that embraces this culture learns and improves by openly identifying and examining

---

[16] The origin of this concept is attributed to Dr. James Reason, professor emeritus at the University of Manchester. Others who have significantly contributed to our understanding of this approach include Sidney Dekker and David Marx. These authors are cited throughout this chapter.

My goal is to provide a *general overview* of this topic as well as some applications and examples. The reader who would like to learn more about this subject is encouraged to consult the list of references at the end of the book; especially those by the authors mentioned above.

process and system weaknesses. In a Just Culture employees feel psychologically safe when voicing any concerns about safety. Human error is viewed as a symptom, not the cause of an adverse event.

A key aspect of a Just Culture is *where* the focus is placed when something unexpected happens. Leaders who ascribe to this belief spend their time asking **what** is responsible, not **who** is responsible. Dekker describes it this way:

> *"The aim of safety work is not to judge people for not doing things safely, but to try to understand why it made sense for people to do what they did – against the background of their engineered and psychological work environment. If it made sense to them, it will for others too. Merely judging them for doing something undesirable is going to pass over that much broader lesson, over the thing that your organization needs to do to learn and improve.*
>
> *What set of circumstances, events, and equipment puts people in a position where an error becomes more likely, and where its discovery and recovery are less likely? The aim is to try to explain why well-intended people can act mistakenly, without necessarily bad intentions, and without purposefully disregarding their duties or safety."*

In a Just Culture three possible behaviors may contribute to an undesirable outcome or an adverse event:

- Human Error
- At-Risk Behavior
- Recklessness

In the next section we will define each behavior. Each of the three scenarios cited earlier will be used to discuss how one of these behaviors was the most likely contributor to the tragic outcome. (I will explain what I mean by *most likely* later in the chapter).

Human Error

The concept of Error Traps was reviewed in Chapter 3. Recall these are situations or circumstances that increase our likelihood for making a mistake. When human error occurs (someone makes a mistake), we define their resulting actions as *inadvertent or unintended.*

When you think about the toddler who was left in a car seat on a summer day, your first reaction is likely, "How could a parent be so absent-minded or careless? There is absolutely no way I would ever make that kind of mistake!"

The following story illustrates how easily this can happen[17].

> *It was Day 3 of a new routine for the Edwards family. Jodie, a professor and counselor at a private university in Cincinnati, had spent the summer of 2008 working two days a week and taking care of her two children: her then 3-year-old son and her 11-month-old daughter, Jenna. On the days Edwards worked, both children stayed with a babysitter near her office.*
>
> *Now it was August and classes were beginning for Edwards, and preschool was starting for her son. Jenna would be with the babysitter Monday through Friday. "I could walk over and see Jenna, nurse her, and bring her back to my office when I wasn't teaching," Edwards says.*
>
> *On Wednesday, August 20, she drove her minivan to her son's Montessori school and took both children inside. "He was really worried about being in a new building, so we went in and stayed with him for 20 minutes, playing and helping him feel comfortable," she recalls.*
>
> *That was the last time the three of them ever played together. Edwards brought Jenna back to the van and strapped her into her rear-facing car seat. "I was talking and singing to her," she recalls. "Five minutes into the drive Jenna started to sing in this little voice she uses when she's sleepy. I had a child-safety mirror, and when I*

---

17 *https://www.parents.com/baby/safety/car/youd-never-forget-your-child-in-the-car-right/*

> *looked in it I could see that she was going to fall asleep." Edwards thought about how much she wanted Jenna to stay asleep and finish her morning nap once she got to the babysitter's. "In a very detailed way, I visualized getting there, walking around to the backseat door, unbuckling her straps, getting her out very gingerly, and covering her ears so the babysitter's door wouldn't wake her. I pictured myself saying to the babysitter, 'Jenna's sleeping. Can I lay her in the crib?'"*
>
> *For the next 15 minutes, Edwards drove toward the babysitter's. But instead of driving past her workplace and traveling another half block to the sitter's house on the next street, she pulled into her office parking lot. "I parked my car," she recalls. "My bags were in the front seat. I walked around and got them out, and I went in to work" -- leaving Jenna in the car on a 92°F day for the next seven hours.*

Let's look at some data on deaths related to this behavior. Since 1998 an average of 37 children have died in hot cars annually.[18] While these incidents are not frequent occurrences, even one child who dies from overheating while in a car is too many.

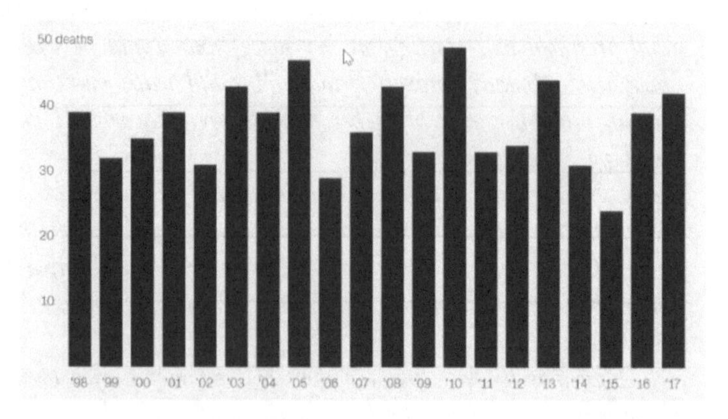

[18] NoHeatStroke.org, a data site run by a member of the Department of Meteorology & Climate Science at San Jose State University, has been collecting data on these incidents since 1998. Graph source: https://www.cnn.com/2018/07/03/health/hot-car-deaths-child-charts-graphs-trnd/index.html

Before we pass judgment on the parents of these children, remember our definition of human error: inadvertent or unintended action. Did these moms and dads **intentionally** leave their children in a hot car for an extended period, knowing the interior of the vehicle would rapidly climb to 130 degrees? No.

Reading through the news articles of these tragic stories, a pattern emerges. The investigations often reveal one or more of these situations contributed to the failure of the parent to retrieve the child from a rear car seat: distraction, time pressure, fatigue, non-routine schedule, or vague guidance (between caregivers). Do these sound familiar? They are Error Traps.

Put aside the legal question about whether this mother was "negligent." That's a discussion for another day and a different book. The sad truth is a loving mother made a mistake. And the price was her infant child's life.

Human error is present every day, everywhere. This includes the workplace. So how do we approach someone who has committed a mistake? In a Just Culture, we don't enter the *Blame Cycle* discussed earlier. We **console** the individual. We approach them with empathy and acknowledge our own fallibility. We seek to understand how we can learn from this mistake. We look for ways to improve the process so we can either eliminate the error or at least make it is less likely to happen again. These improvements often include simple mistake-proofing solutions.[19]

At-Risk Behavior

At-risk behavior can be defined as a choice where the *observed* believes they are in a safe place, but the *observer* judges otherwise.

---

19 *https://www.parents.com/parenting/better-parenting/advice/7-ways-to-not-forget-your-child-in-the-car/*

Using this definition, at-risk behavior is analogous to the adage, "Beauty lies in the eye of the beholder." What is beautiful to you may not be as attractive to me. What you consider unacceptably risky behavior may not seem that way to me. We have may have different perceptions[20] about risk, at least when it involves the specific task I am doing and you are observing me do.

David Marx makes a compelling argument that many at-risk behaviors are based on the way each of us compares the risk we take versus the reward we get (from taking a risk):[21]

> *At risk behaviors exist in many aspects of our lives...For some professionals, such as healthcare providers or airline pilots, at-risk behaviors can be deadly. A pilot may choose not to use a checklist because he believes, after hundreds of successful flights, that he has memorized the steps in a pre-flight check. True, he has them memorized, and every time he mentally recalls the checklist he has a successful flight. Yet his behavioral choice to perform the task by memory is more risky than the decision to pull out the checklist and follow it step-by-step. Perhaps the risk of omitting a critical step goes from one in 10,000 when following the steps line by line (because he could still skip a step), while the risk of skipping a step in recall by memory is one in 1,000.*
>
> *We look in as outsiders, assess the risk more qualitatively than quantitatively, and weigh that increased risk against the utility or benefit associated with the choice. What social utility or advantage does the pilot gain by doing the task by memory? What other task is the pilot able to do when he or she doesn't take the time to do the checklist? We look in as outsiders and judge the conduct of our fellow human beings. Perhaps we say, "Yes, not much value added. I'd skip the step too." Or we observe and decide that the increased risk is not worth the reward. While our friend engaged in the behavior*

---

[20] pp. 30-33

[21] *Whack-a-Mole. The Price We Pay for Expecting Perfection.* David Marx. Your Side Studios. 2009. p35.

may see it differently, we label it "at-risk," a label that implies a difference of opinion on the trade between risk and reward.

Place yourself on an extension ladder on a sunny autumn day. You have set it up on solid footing. You have it leaning at a safe angle. You have a bucket hanging on a rung while you reach into the gutter to remove leaves and debris with your gloved hand. The job goes smoothly as you repeatedly position the ladder, climb up, extract the soggy organic material from the gutter, place it in the bucket, and climb back down.

With your final ascent, you approach the end of the gutter. You can't quite reach the last six inches of the gutter. You are faced with a decision: Do you climb back down and re-position the ladder just to get the last bunch of leaves? Or do you take a bit of a risk by standing on one foot and stretching so your belt buckle is outside of the ladder rail?

What is going through your mind as you make this decision? Have you done these ladder acrobatics before and not fallen? How much risk are you willing to take to save the time required for one more ladder move?

Ask yourself, "What would my wife/brother/friend say if they walked by and observed me contorting my body just to reach a few dead leaves?"

In a Just Culture, how do we react when we see or learn about at-risk behavior? We **coach** people. A coach is in a position to give advice about a better way to do something. And in this case, if someone else believes you are taking an unnecessary risk, you should want them to call you out on your decision *(more on this topic in Chapter 16).*

Remember many times the reason we take risks is because of the risk-taking factors explained earlier. Coaches understand these factors and seek to collaborate with the person taking the risk to mitigate their impact. Once again, the goals are to learn (about why we take unnecessary risks) and to improve (by changing inaccurate perceptions, developing safe habits, and collaborating on making it easier to perform a task).

Reckless Behavior

The following is an excerpt from a recent news story:[22]

> *A man charged with reckless homicide in a crash that killed a teenager was driving at more than double the posted speed limit and was fleeing from another collision less than a mile away, authorities said.*
>
> *The driver was traveling at 107 mph when he slammed into the back of a car carrying a family as it was stopping at a light, officials said. The force of the collision pushed the family's car into the intersection, causing it to crash into another car. The speed limit near the intersection is 40 mph.*
>
> *The teenage girl was pronounced dead a few hours later. Her 12-year-old sister remained in the pediatric ICU with two broken legs, a shattered pelvis, a fractured collarbone and other injuries. She has undergone several transfusions and will need to relearn how to walk. The girl's father suffered spinal and rib fractures and a lacerated spleen and was also still hospitalized.*
>
> *The speeding driver appeared in front of a judge, where his bond was set at $300,000.*
>
> *"It's apparent the motor vehicle he was driving was used as a dangerous weapon," the judge said.*

The third behavior in a Just Culture is defined as the choice to consciously disregard a substantial and unjustifiable risk.

Fortunately, this behavior does not occur nearly as frequently as human error or at-risk behavior. (Although this fact brings no peace to the family in this story whose lives were torn apart). Reckless behavior exists in the workplace. However, it is rarely observed in most industrial environments. Indeed, relative to the other two behaviors, it is a small slice of the pie.

22 *https://www.chicagotribune.com/suburbs/arlington-heights/news/ct-ahp-wheeling-fatal-crash-charge-tl-0802-story.html*

Almost all organizations have established some criteria to discern whether someone is "reckless" or "negligent". The behavior is usually assessed in terms of the degree of being willful or intentional.

What does reckless behavior look like in an industrial setting? Here is one example: a conscious choice to ignore any clearly stated and understood policy which prohibits entering a process where there are extreme hazards without taking certain precautions.[23] If someone has been trained on this policy, understands the consequences of ignoring it, yet consciously decides to violate the policy and unnecessarily exposes themselves or others to substantial hazards, he is reckless.

How do we deal with reckless behavior in a Just Culture? It has to be extinguished. This behavior is **punished.**

Three Behaviors

Let's summarize the behaviors and subsequent actions associated with these behaviors in a Just Culture:

- **Console** human error
- **Coach** at-risk behavior
- **Punish** reckless behavior

} *regardless of outcome*

The point about 'outcome' noted above is important. This means we react to the behavior in the same way, *regardless of whether there was an adverse event associated with that behavior.*

It makes no difference if someone screws up and as a result a bunch of defective product is created or if the same mistake is discovered earlier in the process and there are no quality

---

[23] Some organizations refer to such policies and procedures as LOTO (lock-out-tag-out). Others call it a ZEP process (zero energy potential).

upsets. The person who makes this honest mistake (exhibits human error) is **consoled.**

Imagine a co-worker takes a risk (believing they really were safe) and gets hurt. This is at-risk behavior. On the other hand this individual might be observed taking the same risk, but is not injured. Either way, we would **coach** this person as a first step in moving them toward less risky behavior in the future.

What if an employee exhibits reckless behavior that results in an accident or property damage? Compare this to reckless behavior where no one or nothing is affected. The treatment is the same: some form of **punishment** consistent with local policies designed to strongly discourage this behavior.

Marx proposes this policy statement supporting a Just Culture:[24]

You are a fallible human being, susceptible to human error and behavioral drift. As your employer, we must design systems around you in recognition of that fallibility. When errors do occur, you must raise your hand to allow the organization to learn. When you make a mistake, you will be ***consoled****, and we will collaborate in re-designing the system to minimize errors. When you drift into a risky place, believing that you are still safe, we will* ***coach*** *and collaborate with you in changing the factors which influenced you to take an unnecessary risk. When you knowingly put yourself or others in harm's way, we will take appropriate* ***disciplinary*** *action.*

Two Caveats

We now have a model for a Just Culture in terms of the three behaviors and the action leaders should take when each of these is observed; however, a few words of caution are in order.

First, recognize there will always be "gray areas" when it comes to identifying behaviors. While many times it may seem

---

[24] *Whack-a-Mole. The Price We Pay for Expecting Perfection.* David Marx. Your Side Studios. 2009. p116. (bolded text is my emphasis).

easy to discern which behavior drove the actions of someone, at times it will not be obvious. Recall the example of texting while driving discussed earlier. Both human error AND at-risk behavior were shown as contributors to the potential for an accident. Indeed, some prosecutors argue driving while distracted for any reason could be considered reckless behavior.

Expect differing viewpoints when a group is tasked with assessing an employee's behavior, especially when it involves an adverse outcome. The Just Culture model proposed here is not a recipe for reaching a definitive conclusion on all behaviors. However, leaders who combine this model with sound judgment have an opportunity to strengthen their organization's safety culture by building trust.

The second area of caution is the difficulty in *objectively* looking at the circumstances and actions that contribute to an incident or event. Hindsight bias occurs when people feel they "knew it all along." This bias exists when they believe an event is more predictable *after* it becomes known than it was *before* it became known.[25]

This phenomenon is especially prevalent after outcomes that are exceptionally bad. In these cases, we are much more likely to look for mistakes or behaviors that contributed to the event. The worse the outcome, the more likely we are to see the human errors and poor choices that were part of the causal chain of events. Why is this? Dekker offers a number of possible explanations:[26]

- After an incident it is easy to see where people went wrong or what they should have done.

---

[25] Hindsight Bias. *Perspectives on Psychological Science.* Neal J. Roese, Kathleen D. Vohs. Volume: 7 issue: 5, page(s): 411-426

[26] *Just Culture. Balancing Safety and Accountability.* Sidney Dekker. CRC Press. Second Edition. 2016. pp. 43-44

- With hindsight, it is easy to judge people for missing a piece of data or misunderstanding a communication that turned out to be critical.

- With hindsight, it is much easier to see exactly the kind of harm people should have seen or prevented because it has already occurred.

How can we overcome hindsight bias? Some researchers suggest considering the opposite may be an effective way to get around our cognitive fault. When we are encouraged to consider and explain how outcomes that **didn't** happen **could have happened**, we counteract our usual inclination to throw out information that doesn't fit with our narrative.

In a Just Culture one of the principal questions we should ask ourselves after an incident is, "Did the assessments and actions of the individual *at the time* make sense?" If we can answer this question, it goes a long way toward satisfying the larger goal of a Just Culture as defined earlier:

> *where failure/error is addressed in a manner that promotes learning and improvement while satisfying the need for accountability.*

Ultimately, we want to come up with a plan to reduce the likelihood of this same mistake (or influencing factor) happening again. Accountability is not only about counseling or discipline. It also means a commitment to discovering the true root causes based on what people were thinking at the time they made a fateful decision or took an action which led to a negative outcome.

Now that we have a better understanding of some fundamental safety leadership concepts, we are ready to take a deeper dive into specific approaches that can help us be more effective safety leaders.

PART TWO

The Power of Thoughts

"Everybody is ignorant, only on different subjects"

- Will Rogers

5

YOU CAN'T COACH STUPID

The meeting room was noisy as the construction workers assembled early one morning. The concrete floor and cinder block walls created an echo chamber for the men's voices and the squeaky sounds of metal chairs being pushed into place. Spartan furniture and a dusty, bare floor were tell-tale signs this room was used occasionally for crew meetings, but not much else. Some folding tables on one side of the room strained under the weight of 3-ring binders and manuals stacked halfway to the ceiling. Another table in the back held a large coffee urn and numerous boxes of doughnuts, most of which had already been claimed.

While Russ, the lead trainer for the contractor, went to the front of the room to get the projector running, I settled into a metal chair beside the doughnut table. I fought the urge to grab one of the few remaining sugary treats. Russ planned to talk to the group for the first hour. Then I was scheduled to facilitate the rest of the class.

The class about safety culture and leadership had been requested by the contractor's leadership team. The company recently had some serious injuries and near misses. They were anxious to see what could be done to prevent another event.

The managers were perplexed why some of the guys were taking risks even though they had implored them to "be careful." Even more disturbing, it seemed as though someone was making mistakes on the job almost every week. While all of these were small errors, the senior managers knew any one of these mistakes could cause significant property damage or result in another injury under different circumstances.

As Russ went through the introductory slides and started the first class exercise, a burly man dressed in blue jeans and a shirt that appeared to be one size too small abruptly emerged from an office which adjoined the conference room. I glanced up at the name plate above the opened door. It indicated the stocky man's name was Ed and he was one of the crew leaders.

Ed yanked his office door closed behind him, causing a loud thud when it met the door jamb. I heard him mutter a couple of expletives as he walked briskly by my chair and walked heavily down the stairs on the far side of the room.

Much to my surprise, no one sitting in the metal chairs even turned around to see what all the commotion was about. They acted as if this was a normal occurrence. In the front of the room, Russ cleared his throat and clicked to the next slide in the presentation.

When I turned back to Ed's office, I spotted a large poster taped to the outside of his closed door.

STUPID PEOPLE ARE LIKE GLOW STICKS. I WANT TO SNAP THEM AND SHAKE THE ** OUT OF THEM UNTIL THE LIGHT COMES ON.**

"Hmmm," I thought, "I wonder if this poster reflects his philosophy as a supervisor." I decided to learn more about Ed before jumping to any conclusions.

At the first break, I casually asked several of the guys about Ed. After a bit of hesitation, a few of them opened up.

One fellow took a long drag on his cigarette, looked around to see if anyone was within earshot, and told me in a low voice, "The way you deal with Ed is to keep your head down and don't ask questions. If you're lucky, you can get to your truck and be on the road before he corners you about something you did wrong." The guy beside him shuffled his feet in the dirt and nodded in silent agreement.

At the lunch break, I carefully questioned a few other guys. One response in particular seemed to sum up Ed's approach to people. A tall, lanky man with a full beard and sleeve tattoos offered, "For Ed, there are two kinds of people in the world: Those who can, and those who never will."

Near the end of the class, I decided to talk with Ed. After all, half of the men in this class were on his crew. I needed to hear directly from him. (Ed didn't participate in the class. He explained to Russ he was too busy "putting out fires").

After engaging in some small talk, I asked Ed which guys he thought were his best workers. "That's simple," he said. "I need somebody who shows up on time, takes direction, and doesn't screw up too often. Anything else is a bonus."

Although I expected a response something like the one he gave, I was still taken aback. So I pressed him a bit and asked if he was able to coach anyone who was struggling a little when they were new on the job. "Heck, no! Look, we don't have time to babysit these guys. Either they get it, or they don't. After all, you can't coach stupid."

I had my answer. The poster was not only crude (and tragic) humor. It also reflected the philosophy of the man who taped it to his door!

Ed's phone came to life and vibrated on his desk. He picked it up, looked at me, and motioned me toward the door with his free hand. The conversation ended.

I left his office with a sour feeling in my gut.

Not every person in this organization shared Ed's misguided approach. I met some other supervisors who seemed to be even-keeled in how they led their crews. However, it was clear this supervisory group as a whole could benefit from coaching skills.

Two Mindsets

Ed is an extreme case of someone who has a **fixed mindset**.

Stanford psychologist Carol Dweck introduced the concepts of a fixed mindset and a growth mindset in her seminal book *Mindset.* The differences between these two approaches are stark. In her book, Dweck provides an in-depth look into each of these mindsets.[27]

In general, you have a fixed mindset when you believe your qualities are carved in stone. Your view is you only have a certain amount of intelligence, a defined personality, and a predetermined amount of ability. Further, those who are in positions of authority who share this fixed mindset (teachers, coaches, supervisors, leaders) tend to judge and treat others accordingly. Teachers show favoritism to those with the highest IQs. Coaches give the most opportunities to those with the most innate talent. Fixed mindset leaders expect talented people to perform well and those with less talent to struggle.

On the other hand, a growth mindset is based on the belief your basic qualities (including your intelligence) can be cultivated through your efforts. People who hold this belief embrace the notion everyone can change and grow if they simply apply themselves or are given experiences which help them to learn. This attitude makes a world of difference in the lives we live. Teachers, coaches, and leaders who promote a growth mindset look for ways to challenge their understudies

---

[27] *Mindset. The New Psychology of Success.* Carol Dweck. Ballantine Books. 2008.

to reach their full potential. They don't see themselves or others for who they *are*, but who they *could be*.

A comparison of these two mindsets is given below.

Fixed Mindset	Growth Mindset
Your intelligence is something very basic you cannot change very much	No matter how much intelligence you have, you can always change it quite a bit
You can learn new things, but you really can't change how intelligent you are	You can always do something to substantially change how intelligent you are
You are a certain kind of person and there isn't much you can do to change this	No matter what kind of person you are, you can always change substantially
You can do things differently, but the important parts of who you are can't be changed	You can always change the basic things about the person you are

Fortunately, the growth mindset can be taught to managers. Researcher Peter Heslin and his colleagues proved this by designing and conducting a workshop with this goal.[28]

The workshop starts by helping people understand how the brain is dynamic. It shows how the brain changes with learning and claims just about anyone can develop new abilities, regardless of their age, with coaching and practice. The workshop also includes exercises where the participants:

---

[28] Keen to Help? *Managers' IPT and Their Subsequent Employee Coaching.* Heslin, VandeWalle, and Latham. Personnel Psychology 59 (2006) 871-902.

- learn why it is important to understand people can develop their abilities
- think of personal examples of when they started with low ability, but now perform at a higher level
- write to a protégé who is struggling with some encouragement
- reflect on a time when they saw someone learn to do something they never thought they could do.

Interestingly, the participating managers were much more willing to coach a poor performer and were more likely to provide coaching suggestions. These traits persisted many weeks after the workshop concluded.

There are several implications of this work. The workshop creators conclude a growth-mindset environment can be created that will yield huge dividends in terms of leveraging the full potential of all employees, regardless of their innate skills. In order to foster such an environment, leaders need to –

- believe all skills can be learned
- communicate that the organization values learning and perseverance (not just ready-made genius or talent)
- provide feedback that promotes future learning
- insist managers are available as key resources for learning, not just directing work.

Let's go back to the story about Ed. In his world, people either had the "right stuff" or not. He dismissed the belief you can coach people to a higher level of performance, which is a hallmark of someone with a growth mindset.

Supervisors or managers with a fixed mindset severely limit the progress of the organization. Employees whom they supervise tend to be disengaged. Turnover is high. Trust is minimal. Success is measured in absolutes. And perhaps most importantly, learning and improvement are crushed.

This combination of outcomes from a fixed mindset approach makes it impossible to foster a Just Culture, which was discussed in the last chapter. A Just Culture is the foundation for achieving an environment centered on commitment.

Is it any wonder this company struggled with their safety performance?

When I talked to one of the senior managers, I learned more about Ed. I was told his promotion as a supervisor over 10 years ago was largely based on his work ethic and being able to get the job done. He was considered to be their "ace" troubleshooter. He measured everyone else against these standards. Training sessions for supervisors were almost exclusively directed toward improving their technical skills. Neither Ed nor any of his peers were provided with any coaching skill development.

What should be done? One manager confided Ed was considered a true subject matter expert. Many of the other supervisors came to Ed for advice – not about people (thankfully), but how to solve a particularly thorny or difficult construction problem.

I recommended Ed be given the opportunity to learn how to coach others. This could not be limited to a few hours of advice. It needed to include someone to mentor Ed in applying and practicing coaching skills. He would need immediate and candid feedback.

Further, I advised if Ed did not demonstrate a willingness to learn and implement basic coaching as part of the way he supervised others, then the manager would have a choice. He could allow Ed to stay in his current role and poison the organization. Or he could make the hard decision to find another position where they can benefit from Ed's expertise – a role which had no direct supervisory responsibility.

We don't want to just give up on people like Ed. If we did, then we could rightly be accused of harboring the same fixed mindset we want to change! However, we should not tolerate behaviors associated with an extreme fixed mindset. Anyone in a significant influencing role should know how to use basic coaching skills. Or they need to find another line of work.

It's the only way for an organization to grow and provide a safe work environment – both physically and emotionally.

Conclusion

Reflect on your own mindset. Do you believe you can learn new things, but you really can't change how intelligent you are? Do you believe you are a certain kind of person and there is not much you can do to change this? Do you value existing talent over learning? If so, you have a fixed mindset.

The first step toward realizing your full potential is an honest self-assessment. If you have a fixed mindset, develop a plan to move toward a growth mindset by deploying some of the strategies proposed by Dweck, Heslin, and others.

If you are in a leadership role, take what you learn and create an environment that nourishes a growth mindset within your organization. The results may surpass your expectations.

- **Perform an honest self-assessment on your personal mindset.** The first step to getting the most out of yourself and others is to recognize everyone (including you) has the capacity to learn and grow. Choose an endeavor where you are not an expert and challenge yourself to learn something. Seek constructive criticism from your peers.
- **Provide opportunities for people in the shadows.** In safety discussions, do you presume only a few people are capable of contributing? Try to purposefully involve people in safety activities who may not have been asked to participate in the past – even if it is in a small way. You might be pleasantly surprised at how they step up to challenge.
- **Set the expectation with your direct reports they need to develop their employees.** One performance measure could be how many of their employees are engaged in things that matter. If your supervisors do not have the skills to develop others, you may need to "coach the coaches."
- **Role model the behavior you expect.** Schedule time with your direct reports and discuss opportunities for them to learn and grow personally and professionally. Publicly admit your own mistakes, so others see you as fallible.
- **Recognize and/or reward persistence and grit.** This is especially true if the individual does not initially achieve the desired outcome. Provide feedback that encourages more effort.

- **Be thoughtful about how you react to human error.** Emphasize the importance of learning from mistakes and collaborating to improve the process in a way that minimizes the likelihood of future mistakes. Lead in a way that creates and sustains a Trust Cycle.

"Normal people live distracted, rarely fully present. Weird people silence the distractions and remain fully in the moment."

\- Craig Groeschel

6

YOUR DANGEROUS & DISTRACTED MIND

Distractions are everywhere in our world. We can be distracted while driving, while working, or while doing any number of routine tasks. In the workplace, incident investigations reveal tens of thousands of injuries each year occur when people are not focused on the task at hand.

Most everyone recognizes the dangers associated with being distracted while driving a motor vehicle. Distracted driving is a leading contributor to automobile crashes. For example, here are just a few statistics from several organizations dedicated to stopping texting and driving injuries and deaths:

- Every year, about 421,000 people are injured in crashes involving a driver who was distracted in some way.

- Each year, over 330,000 accidents caused by texting while driving lead to severe injuries. This means over 78% of all distracted drivers are distracted because they have been texting while driving.

- 1 out of 4 car accidents in the US are caused by texting while driving.

- Texting and driving is 6 times more likely to get you in an accident than drunk driving.
- It takes an average of three seconds after a driver's mind is taken off the road for any accident to occur.
- Reading a text message while driving distracts a driver for a minimum of five seconds each time.

Error Rate and Distraction

A recent Michigan State University study[29] provides supporting evidence being distracted significantly increases human error (which can result in an accident).

Participants in this study were asked to perform a series of tasks in order, such as identifying with a keystroke whether a letter was closer to the beginning or end of the alphabet. Of course, a certain number of errors were made even without interruptions.

Occasionally the participants were told to input two unrelated letters -- which took about 3 seconds -- before returning to their task. These slight interruptions led to participants making twice as many mistakes when they returned to their sequencing task.

In addition, there are a number of studies and/or exercises which prove humans cannot consciously complete more than one task at a time. Indeed, one author has labeled the so-called skill of multi-tasking as "worse than a lie." No matter how you look at it, being distracted (for any reason) significantly increases the risk of making a mistake and/or being injured.

Unfortunately, we are not only distracted by something in our environment, but simply by the way our minds operate!

---

[29] *Momentary interruptions can derail the train of thought.* Altmann, E. M., Trafton, J. G., & Hambrick, D. Z. Journal of Experimental Psychology: General, 143(1), 215-226.

Subconscious Distraction

A fascinating study conducted by Matthew Killingsworth and Daniel Gilbert provides insight into how frequently our minds wander.[30] The authors point out humans spend a lot of time thinking about what is not going on around us. Further, we often contemplate events that happened in the past, might happen in the future, or will never happen at all. As part of their work, Killingsworth and Gilbert devised a clever way to measure how often we think about something other than what is happening right here and now.

The researchers needed to find an efficient way to measure real-world emotions. They required a process that would allow them to contact people as they engage in their everyday activities and ask them to report their thoughts, feelings, and actions at that moment.

The problem was solved by developing a web application, which was used to create a large database of real-time reports of thoughts, feelings, and actions of a broad range of people as they went about their daily activities. The application contacted participants through the phones at random moments during their waking hours, presented them with questions, and recorded their answers to a database. The database currently contains nearly a quarter of a million samples from about 5000 people from 83 different countries who range in age from 18 to 88 and who collectively represent every one of 86 major occupational categories.

To find out how often people's minds wander, what topics they wander to, and how those wanderings affect their happiness, they analyzed samples from 2250 adults who were randomly assigned to answer a happiness question ("How are you feeling right now?"), an activity question ("What are you doing right now?"), and a mind-wandering question ("Are you

---

30 *A wandering mind is an unhappy mind.* Killingsworth, M. A., & Gilbert, D. T. (2010). Science, 330(6006), 932-932.

thinking about something other than what you're currently doing?").

In terms of the third question, here was the key finding: people's minds wandered frequently, regardless of what they were doing. How frequently? Mind wandering occurred 47% of the time. Surprisingly, the nature of people's activities had only a modest impact on whether their minds wandered.[31]

In other words, nearly half the time we are not truly focused on what we are doing at the moment!

Mind on Task

In the workplace, consider the classic admonition about "keeping your mind on the task." A supervisor who strives to reduce risk will collaborate with employees to remove all obvious distractions so they can focus on the task at hand.

However, armed with the mind-wandering information cited above, the challenge to keep our minds on our task goes beyond removing common distractions. Even if we eliminate the most egregious distractions (cell phones, pagers, interruptions, environmental factors, priority changes, scheduling demands, etc.), there is a good chance people still may not think about what they are doing when they do it.

In other words, what we can't see when observing others is an insidious source of risk. Do we really know what others are thinking about when they are doing a job? Of course not.

Therefore, we should assume **50% of the time, anyone who is performing an activity is not totally focused on what he is doing.**

Now that's a scary thought!

---

[31] The study also revealed some interesting results from the answers to the first two questions about happiness and activity. The findings of those aspects obtained from the self-reported data are not discussed here.

Conclusion

There are steps we can take to reduce the risk associated with a wandering mind. These include taking frequent time-outs, developing safe habits, and using social influence to encourage everyone to speak up. Additional recommendations are listed in the **SAFETY LEADER'S TOOLBOX™** on the next two pages.

A creative and thoughtful mind is beautiful. A wandering mind that is not focused on a high risk task is dangerous!

- **Eliminate obvious distractions.** You likely know the kinds of activities that could prevent someone from keeping his full attention on the task at hand. It could include a behavior like using a cell phone on the operating floor. Educate the workforce on the dangers of operating equipment or working while distracted. Write a policy that includes what it is AND *why* it is important. Provide positive reinforcement for those who abide by the policy. And enforce it with those who do not.
- **Assess whether you are creating distractions.** You may unintentionally cause distractions by your leadership actions. If you are constantly changing priorities or interrupting others while they are working on a task, you are not setting a good example.
- **Address tasks with the highest risk.** The jobs or tasks with the greatest risk for injury should be prioritized for action. It is far more important to intervene when the task has a high risk profile and therefore demands a person's full attention to safely complete this task.
- **Have a proactive conversation.** The only way to know what someone is thinking about is to have a conversation. Ask them to share why they are doing what they are doing. We may discover false perceptions or risk-taking habits we can proactively address. Of course, this approach is only effective if the conversation is facilitated in a way that builds trust, learning, and improvement. *(see Chapter 21)*.

- **Take time-outs.** Explicitly give your employees permission to take a brief "time-out" at frequent intervals when they are working on a high-risk task. Brief time-outs are different from most interruptions or distractions. These are *planned personal pauses* designed to keep our minds "in the moment." Time-outs are employed to become more mindful about your surroundings and to regain focus on the task.
- **Develop safe habits.** One of the most effective defenses for a wandering mind is to practice performing a task the same (safe) way every time until it becomes a habit. When someone loses focus, a safe habit is how this person will subconsciously perform the task. Habits allow individuals to safely complete a task, even if they are temporarily distracted.
- **Leverage social influence.** Enlist respected employees in the organization to set the standard for remaining focused while performing high-risk work. When opinion leaders (1) frequently use time outs and (2) constantly work on developing safe habits, many others will follow their lead. *(see Chapter 15).*
- **Empower others to speak up.** Employees always need to watch out for one another. There are a series of actions leaders can take which will encourage others to say something if they see risky behavior or if they perceive a co-worker is not completely focused on performing a critical task. *(see Chapter 17).*

"People put a lot less effort into picking apart evidence that confirms what they already believe."

- Peter Watts

7

I SEE WHAT I EXPECT TO SEE

Confirmation bias is the tendency to process information by looking for or interpreting information consistent with one's existing beliefs. This biased approach to decision-making is largely unintentional and often results in ignoring inconsistent information.[32]

Existing beliefs can include one's expectations in a given situation and predictions about a particular outcome. People are especially likely to process information to support their own beliefs when the issue is highly important or self-relevant.

In Chapter 1 the differences between two distinct cultures (compliance and commitment) were highlighted. This chapter describes how a confirmation bias can perpetuate a culture of compliance. I will also discuss how the conversations which take place in a workplace with a *culture of commitment* minimize the potential for confirmation bias.

---

[32] Photo License: Creative Commons 3 – CC BY-SA 3.0. Attribution: Nick Youngson

Culture of Compliance

The model below explains how confirmation bias can influence decision-making (and the actions taken by managers) when an organization is managed through compliance.

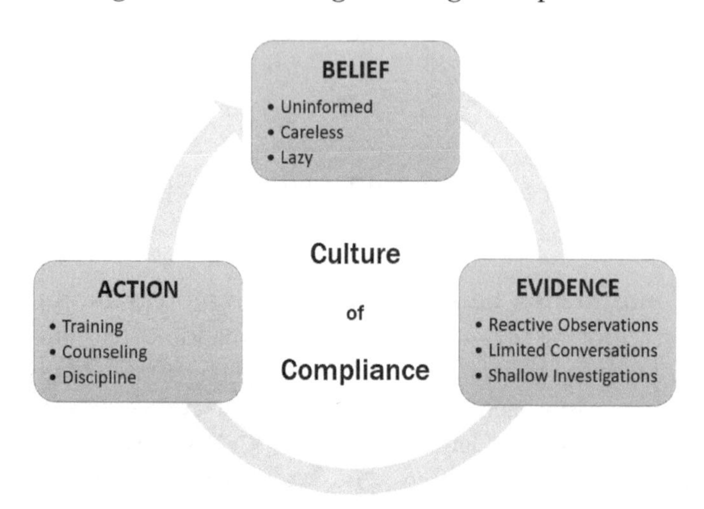

"Why do people violate rules or take unnecessary risks?"

It begins with a person's existing beliefs.

Supervisors and managers know one reason a person could violate a rule, policy, or procedure is if they are uninformed (because of poor training or lack of experience).

However, if a subsequent conversation with this employee reveals he has done this specific task many times before and/or has been trained, a supervisor who is part of a culture of compliance often shifts to another explanation. He concludes the reason this person ignored a rule, took an unnecessary risk, or made a mistake is because of poor work habits. For example, the employee might be labeled as careless, inattentive, or lazy.

By reaching this conclusion, the supervisor could be making a **fundamental attribution error.**[33] Indeed, many of us are prone to making this error because we often attribute people's behavior to the way they are, rather than to their situation.

Confirmation bias emerges when the supervisor seeks to validate his existing beliefs by looking for supporting evidence. When seeking this information, the supervisor will tend to look for evidence that confirms his conclusion is true, rather than pursue information that could prove his view is inaccurate. In a culture of compliance, the supervisor obtains this supporting evidence through mostly cursory methods. These include:

- making observations only after an event has occurred
- initiating brief, one-sided conversations with employees that are mostly directive in nature
- conducting "check-the-box" incident investigations that fail to uncover root causes

The action taken by the supervisor is contingent upon the evidence he or she obtains. If his or her assessment reveals a genuine lack of experience or knowledge has resulted in a rule or policy violation, the employee will be given the appropriate job training.

But what happens if the individual has received sufficient training? Or he has performed the task successfully many times? And yet this person still does not follow a rule, policy, or procedure (or takes an unnecessary risk)? In a culture of compliance, the supervisor's belief system leads him to make a decision that is frequently ill-advised. He reasons since the employee knows how to do the task, he must be careless, lazy, or inattentive. Therefore, the only way to change this unacceptable behavior is through counseling or discipline.

---

[33] p. 48

By taking this action, the supervisor sincerely believes he is fulfilling his obligation of "holding people accountable". After all (he believes), conscientious employees who are sufficiently trained would never violate a rule, policy, or procedure.

Likewise, any reasonable individual certainly would not take an unnecessary risk. Armed with these beliefs, the supervisor expects if employees are trained, only those with a poor work ethic (careless, lazy, inattentive, etc.) will violate a rule or take a risk.[34]

Indeed, my belief system in this scenario is continuously reinforced because of my confirmation bias:

1. I believe some people could violate a rule because they are uninformed. However, most rule-breakers and risk-takers are either lazy or careless.

2. My "evidence" is based upon the logic that if a person is trained and given clear expectations, there is no reason for them to violate a rule or take an unnecessary risk.

3. I address most non-compliance through warnings or counseling.

4. Assuming these persons have received training, the next (lazy) rule-breaker or (careless) risk taker I encounter will need discipline to correct this behavior.

One way to break this cycle is to look for information (evidence) in a way that objectively considers alternative reasons for risk-taking or errors. We will examine this approach next.

---

[34] People may violate a rule or take a risk for reasons that have nothing to do with personal attributes. Social influences, procedural drift, vague guidance, and multi-tasking are just a few factors to consider.

Culture of Commitment

Let's look at the decision-making model prevalent in a culture of commitment. Supervisors who are part of a culture of commitment also hold certain beliefs. However, many of these beliefs are notably different when compared to those in a culture of compliance. Why? They have been developed through a set of experiences unique to a culture of commitment. These beliefs are rooted in a deeper understanding of human fallibility as well as our susceptibility to risk-taking.

Regardless of the culture in which they work, most

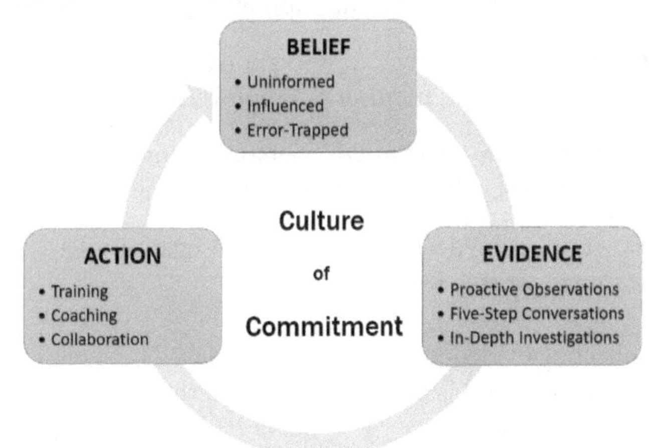

"Why do people violate rules or take unnecessary risks?"

supervisors actually share a common belief. They (correctly) believe a primary reason anyone could potentially violate a rule, policy, or procedure is the employee is uninformed.

But this is where the two belief systems diverge.

In a culture of commitment, if it is confirmed the employee is knowledgeable and has been trained to perform the task, the supervisor's experience yields a different set of beliefs regarding these undesired behaviors. He realizes most of the time if an employee ignores a rule, takes an unnecessary risk, or makes an error it is because the worker is:

- subjected to a risk-taking factor
- set up to make a mistake (because of a situation referred to as an error trap)

The supervisor will seek to validate his beliefs by looking for evidence. However, unlike someone who manages through compliance, the supervisor who fosters a culture of commitment is more diligent and objective in his search for evidence. As a result, confirmation bias is much less likely to occur.

Methods used to obtain the supporting evidence include:

- making proactive observations to understand the situations and mindset of employees
- facilitating structured, personal conversations with employees that encourage shared learning
- conducting detailed incident investigations designed to uncover true root causes

Collecting evidence in this way is crucial. These discoveries frequently support his belief system about the causes for errors and/or risk-taking.

As before, the action taken by the supervisor is contingent upon the evidence he or she obtains. If he uncovers a lack of experience or understanding by the employee, further training on how to perform the specific tasks of the job is warranted.

However, what if the individual is adequately trained? In a culture of commitment, the supervisor knows over 90% of human error[35] and many at-risk behaviors are a result of process or organizational flaws.

The supervisor may decide to do any number of things. All the actions listed below are preceded by a sincere, caring

---

[35] *Out of the Crisis.* W. Edwards Deming. Massachusetts Institute of Technology. Cambridge, MA. 1986.

conversation with the employee where an assessment is made to determine the root cause:

- The employee could have decided to take a risk or violate a rule because of a false perception or a risk-taking habit. In this case, the supervisor will engage in coaching or counseling.
- Perhaps the employee decided to take a risk or violate a rule because extra effort was required or resources were unavailable. If so, the supervisor will seek to collaborate with the employee on making it easier to perform the task safely..
- When someone makes a mistake, the supervisor will look for potential error traps. He will mitigate these situations or implement mistake-proofing solutions with input from the employee.

These actions reinforce the commonly held belief system in a culture of commitment that employees are often in situations that encourage them to take risks or they are set up to make mistakes.[36]

To summarize this decision-making model: A virtuous cycle is created where employees are given the "benefit of doubt" on why they violated a rule or had taken a risk. This is the foundation of the supervisor's belief system. The evidence nearly always reveals the root cause as risk-taking factors and/or error traps. The actions taken reflect a genuine caring for the employee as well as a desire for learning and improvement.

To compare and contrast how these decision-making models work, let's look at a case study.

The following story is based on an actual event.

---

[36] In a Just Culture, if a supervisor determines that the employee made a *conscious* choice to disregard a *substantial* and *unjustifiable* risk, this behavior warrants some kind of discipline.

The Case of the Missing Gloves

One of the work rules at a large food processor is employees are required to wear latex gloves whenever handling any ingredients. This is essential not only to protect the employees' hands, but as a hygiene and food safety requirement.

Although every new hire was informed about this rule and gloves were available, some employees were occasionally observed not wearing gloves. To encourage compliance, supervisors issued verbal or written warnings. But the problem persisted.

At the same time, the plant was faced with periodic episodes of plastic contamination in their food products. The employees were instructed to be more attentive when processing the raw materials. They were told to be sure plastic was not allowed to enter the production system. Moreover, it was made clear to the employees disciplinary action would be taken if any plastic entered the production process in their work area.

It turns out these two events were related. The connection was made when an analysis of the plastic contaminants in the food showed some were made of latex. The color of these foreign materials also matched the color of the gloves.

Employee interviews revealed an astonishing contradiction. The workers knew they were required to either wear the gloves or receive a reprimand – that much was clear. However, they told the interviewers if plastic of any kind was found in their production line, they could also be disciplined. The punishment for a "plastic contaminant violation" was more severe than for "glove non-compliance" because the financial losses associated with contamination were very large.

In addition, there was only one size of glove (XL) available for everyone. And the smaller employees (mostly women) could not wear these gloves because they would often fall off! It was a true conundrum.

What did these employees do? They assessed which policy or rule to violate which would minimize the severity of any potential discipline. It was no surprise they often chose not to wear the gloves when they thought no one was looking. (They would wear the over-sized gloves when a supervisor was present, then remove them after he left the area).

In this culture of compliance, the area supervisor held a belief the reason these women did not wear gloves was they were careless, lazy, inattentive, etc. He used confirmation bias to gather evidence through brief, directive conversations. Because his cursory investigation didn't uncover a reason for the employees not to wear the gloves (they knew the rule and the gloves were available), the supervisor took action by either counseling them or issuing discipline.

In the decision-making cycle for a culture of commitment, there would have been a much different outcome. In this situation, the supervisor would have realized people violate rules for a number of reasons. A deeper conversation (initiated because he cared about his employees) would have almost certainly revealed ill-fitting gloves as a legitimate reason this policy was violated. The supervisor may have also discovered a number of error traps which could contribute to the mistake of gloves ending up in the process (e.g., rushing, frustration, the environment). His actions would have likely centered on collaboration with the employees to address these flaws.

In this case, the solution is obvious. Provide gloves in enough sizes to fit everyone comfortably so they can do their jobs. However, these kinds of positive outcomes are only possible if supervisors engage their employees in dialogues free from confirmation bias.

Conclusion

What kind of conversations are you having with your employees? Unless your conversations are rooted in caring (and they are conducted in a manner consistent with a culture of commitment), you are susceptible to confirmation bias. This unintentional decision making can keep you in a state of managing for compliance - where the principal objective is to maintain or exert control.

To foster learning and improvement, you need to ask questions (and find answers) that reveal the **real** reasons why people violate rules or take risks. Do you have the courage to look for evidence or gather information that could challenge your existing belief system?

It starts with being sincere and diligent in the conversations you have with others.

- **Look for what people are doing right.** If you look for things that are wrong, you are more likely to find these flaws. If you look for safe behaviors, you will have ample opportunities to observe these as well. Provide positive reinforcement for any behaviors you would like to see repeated *(see Chapter 12)*.
- **Don't jump to conclusions.** If you judge someone too quickly for why they took a risk, you will almost certainly be mistaken. That's why it is critical to conduct the right kind of safety conversation. When facilitated effectively, a conversation uncovers potential risk-taking factors and error traps. These are often the precursors to risk-taking behaviors or events *(see Chapter 21)*.
- **Conduct detailed incident investigations**. It takes less effort to complete an incident investigation by only gathering the information required by the form. A comprehensive assessment takes more time, but it includes thoughtful inquiries and focuses on the employee's frame of mind and the circumstances that may have contributed to the event. Be aware of your own biases which could influence how you see the evidence and any subsequent conclusions.

"Anything that just costs money is cheap."

\- John Steinbeck

8

MY STUFF IS WORTH MORE

Humans are complicated. While some of our base emotions and behaviors are easy to understand, there are times when we appear to make irrational decisions when faced with personal change. For example, behavioral economists have identified a specific instance when we apparently place a very different value on something depending upon whether we own it or not. Consider the following scenario.

Suppose a team performed an analysis on the layout of a work area. The team concluded a significant amount of waste of motion and transportation would be removed if the work stations in the cell were re-arranged. With a proposed new floor layout, each of the operators would walk shorter distances as they moved among the stations. It would make it easier for them to accomplish their work each day. The location of the new work stations would be comparable in every way to the existing work stations – tools, space, lighting, climate, proximity to the work. This sounds like a positive outcome for everyone!

However, when the proposed plan was shared with the crew, it was met with surprising resistance by some of the operators. This would seem to be an illogical decision. The operators would rather walk further (and therefore work harder) than accept these minor personal changes to their work flow! How can this be?

We could simply attribute this response to resistance to change. While at a high level this might be true, there could be a more specific reason for their reluctance.

Endowment Effect

The endowment effect is a cognitive bias which was first hypothesized by economist Richard Thaler. According to Thaler's theory, people value something more if their ownership is clearly established. In fact, people often demand much more to sell an object than they would be willing to pay to buy it.

Numerous studies have been conducted to explore the endowment effect.

In a classic study first conducted by Daniel Kahneman and others,[37] people were asked to assess the value of a coffee cup which had been given to each of them. Specifically, they were asked "How much would you be willing to sell your coffee cup for?" A second group in the study was asked how much they would be willing to pay for the same coffee cup. But this group did not own their coffee cups – they were only shown the cup and asked to value it.

The results? The subjects who owned their coffee cups consistently valued them higher (about 2x) than the value given to the same cups by the group of non-owners. In some cases, the first group even said they would prefer to keep their coffee cups rather than selling them.

The basic conclusion from this research is *people prefer things with which they have been endowed (given ownership).*

---

[37] Kahneman, Daniel; Knetsch, Jack L.; Thaler, Richard H. (1990). *Experimental Tests of the Endowment Effect and the Coase Theorem.* Journal of Political Economy. 98 (6): 1325–1348.

IKEA® Effect

A variant of this concept has been shown to increase the perceived value of something even more than ownership. The researchers who initially demonstrated this phenomenon coined the term "IKEA® effect". This name is derived from the Swedish manufacturer whose products typically require some assembly.[38]

A series of studies was conducted where participants were asked to either build simple boxes, fold origami, or construct various LEGO® models.[39]

In the initial experiment, participants were paid $5 to participate. Some were randomly given the task to assemble a plain black IKEA® storage box. They were given an unassembled box with the assembly instructions included with the product.

The other participants were given a fully assembled box and provided with an opportunity to inspect it. After the boxes were assembled by the first group, each participant was asked to make a price bid on the specific box they had either built or were assigned to inspect. (All boxes, whether built by the participants or already assembled, were identical).

The results showed "builders" bid significantly more for their boxes than "non-builders." Thus, while both groups were given the chance to buy the same product, those who assembled their own box valued it more than those who were given the chance to buy an identical pre-assembled box. Interestingly, the researchers also solicited the two groups for how much they liked the boxes they either built or were assigned. A similar effect was observed, with builders reporting a higher degree of liking for their box compared to non-builders.

The researchers followed this experiment with another one. This time they asked some participants to create either an

---

[38] IKEA® is a registered trademark of Inter IKEA Systems B.V.

[39] LEGO® is a trademark of the LEGO Group of companies.

origami frog or crane. The participants were then offered a chance to buy their creations with their own money. This second experiment was designed to benchmark the magnitude of the IKEA® effect by comparing participants' willingness-to-pay for their own creations to two different standards.

First, a different set of participants was asked to bid on the builders' origami. The purpose was to determine how far above the market price the builders priced their own creations. Second, some experts were asked to make the same origami. The researchers then solicited bids for these professional creations. This allowed a comparison in the values the novice participants placed on their creations to the same origami made by experts.

The results showed the builders' valuation of their origami was over *four times more* than non-builders were willing to pay for these creations. Thus, while the non-builders saw the amateurish creations as nearly worthless, the builders assigned their origami significant value. Additionally, builders were willing to pay nearly as much for their own creations as the neutral set of non-builders were willing to pay for the well-crafted origami made by the experts!

The experiments described above demonstrate the significance ownership or our involvement in the creation of something has on the value we place on things.

With this background, let's revisit our scenario about the proposed work cell layout change. If these operators had been working within the existing layout for an extended period of time, it is likely they had a sense of ownership of their work space. "I do my work right here. I *own* this space." The endowment effect suggests these individuals will look for a significant increase in value if they are asked to move to a new location – even if it is comparable in every way.

In fact, these operators almost certainly "created" or modified some aspects of their work space. These could be simple things like customized places to hang their tools or store their lunches. Perhaps they installed a display board with personal mementos. Any of these small design changes would not only increase their sense of ownership, but also support a feeling they helped to build their current work cell.

In this light, their behavior is not irrational. Each person is looking to offset the pain associated with the anticipated loss of their personal space. They would rather stay where they are to avoid this loss. Kahneman terms this "loss aversion", which is the human tendency to strongly prefer avoiding a loss to receiving a gain. He suggests loss aversion is the primary reason people exhibit the endowment effect.

So what are the implications for you as a leader? Two situations come to mind:

Change Management

Managing through personal change can be difficult. If we appreciate the endowment effect, it gives us greater insight into why an individual's resistance may be so strong. When loss aversion is in play, it will take something of significantly greater value than the pain of their loss to enable change. In other words, the classic phrase of WIIFM (What's In It For Me?) has a whole new meaning.

The person who is being asked to change has a value in mind twice as great as the perceived loss they are asked to incur. We need to answer an individual's WIIFM in a way that overcomes their loss aversion. What can we tell someone about the benefits of the change that will be of sufficient value to overcompensate them for their pending loss? To increase the likelihood of acceptance, one strategy could be to enlist an *opinion leader* to be the spokesperson for change.[40]

Project Management

Have you noticed some project leaders struggle with canceling a project, even when it is clear this is the best decision? Some projects that should have been completed in six months can

---

[40] *Opinion Leaders* are discussed on pp. 160-162

drag on for a year or longer. The leader uses various reasons to keep the project active. While some are legitimate, many are just excuses to delay the inevitable.

> *"I just need one more piece of data..."*
> *"As soon as we get some time to..."*
> *"We are close to a solution, just one or two more trials..."*

It could be the project leader is reluctant to admit failure because he/she doesn't want to experience the pain of losing. In this case, we need to explain how the value of closing the project is greater than keeping it open indefinitely.

Another implication of the endowment and IKEA® effects is the notion any idea that does not originate from our work circle is likely to be inferior. Perhaps you have heard this referred to as the "not invented here" syndrome. As we can see from the psychology of ownership and creation, we are prone to overvalue our own creations (ideas) and by default undervalue the creations (ideas) of others.

Conclusion

When people own (or have a strong attachment to) something, it is more difficult for them to give this up than if they do not own it. This so-called "endowment effect" can make it more difficult for people to accept personal change or to move on to something else if they experience a loss in performance or value.

Equally important is the IKEA® effect, where our ideas are highly valued if we were the originators. At the same time, we tend to undervalue ideas that come from someone with whom we do not have a relationship.

We should consider these phenomena when trying to influence others. The perceived value of any incentives may need to be significantly higher than the value of the status quo in order to effect the desired change.

- **Give employees explicit ownership.** The possibilities are endless. In a safety context, here are some examples:
    - *Make it clear they own their personal safety.* They are empowered to stop any process or task if they feel there is a risk of someone getting hurt.
    - *Make PPE personal.* Offer to provide custom-fitting gloves, earplugs, face shields, chemical suits, etc. Let them choose their style/size of PPE which meets the job requirements. They are more likely to wear these items if they are "owned."
    - *Assign ownership of operating areas.* This is especially helpful for housekeeping or 5S.[41] If several people share the same work area across multiple shifts, consider segregating the area into specific areas of responsibility. Hold people accountable for cleaning/auditing/maintaining the section they own.

---

[41] 5S is a workplace organization process. The Five S methodology helps a workplace remove items no longer needed *(sort)*, organize the items to optimize efficiency and flow *(straighten)*, clean the area in order to more easily identify problems *(shine)*, implement color coding and labels to stay consistent with other areas *(standardize)* and develop behaviors that keep the workplace organized over the long term *(sustain)*.

- **Engage employees in generating safety solutions.** This is the IKEA® effect. If the idea is theirs, it is more likely to be adopted because it is likely to be highly valued. This is an important consideration when implementing a process change intended to address a safety hazard or mitigate a potential risk. You could take the approach of letting everyone know, "this is the way you will perform this task from now on." But how well would this change be accepted?

  Which would you rather have?

  a) A solution that is 90% effective in mitigating the risk, but is unlikely to be adopted by the work force?

  b) A solution that is 70% effective in risk reduction, but is embraced by the people who do the work (because it was their idea)?

"Diligence is the mother of good luck."

- Benjamin Franklin

9

I'M JUST UNLUCKY

Perhaps you know of someone who considers most things that happen are caused by chance. Or they seem to accept the outcome of whatever significant events occur in their life was determined by fate or luck.

On the other hand, some people clearly believe they control their own destiny. Their belief is whatever happens to them is mostly due to the choices they make or the actions they take.

These disparate belief systems represent opposite ends of a continuum social scientists refer to as **Locus of Control.** A person's Locus of Control (LOC) is where someone places the primary causation of events in his life. Those who believe their life is largely controlled by outside forces (externals) are on one end of the spectrum. Those who believe they control their own lives (internals) are on the other end of the spectrum.

Locus of Control is a psychological construct. This simply means it is an instrument that can be used to describe a group of attitudes or behaviors.

Julian Rotter is credited with introducing the concept of Locus of Control. He based much of his research on the work of Albert Bandura, who developed social learning theory. In his seminal paper published in 1966, Rotter explains people can interpret events as being either a result of one's own actions or external factors.[42] Rotter developed a scale to assess whether a person has a tendency to think situations and events are under their own control (internal influences) or under the control of someone or something beyond their control (external influences).

Listed below are a few of the paired statements from the original *Rotter Internal-External Locus of Control Scale:*

> **1a.** Becoming a success is a matter of hard work; luck has little or nothing to do with it.
> **1b.** Getting a good job depends mainly on being in the right place at the right time.
>
> **2a.** Many of the unhappy things in people's lives are partly due to bad luck.
> **2b.** People's misfortunes result from the mistakes they make.
>
> **3a.** One of the major reasons we have wars is because people don't take enough interest in politics.
> **3b.** There will always be wars, no matter how hard people try to prevent them.
>
> **4a.** In the long run, people get the respect they deserve in this world.
> **4b.** An individual's worth often passes unrecognized no matter how hard he tries.
>
> **5a.** Many times I feel I have little influence over the things that happen to me.

---

[42] J.B. Rotter. (1966). *Generalized expectancies for internal versus external control of reinforcement.* Psychological Monographs, 80, (1, Whole No. 609).

5b. It is impossible for me to believe chance or luck plays an important role in my life.

Since Rotter's original work, many other researchers have studied Locus of Control, often seeking to determine if it can be used to predict outcomes in more targeted domains. Here are a few examples of Locus of Control scales developed for specific purposes:

- The *Multidimensional Health Locus of Control Scale* is used to assess an individual's belief in what influences their health.
- The *Drinking Locus of Control Scale* is focused on alcoholics and those who regularly consume alcohol to assess whether the person believes they can control their drinking.
- The *Parental Health Belief Scales* are used to assess the extent to which a parent believes they have control over their child's health.
- The *Traffic Locus of Control Scale* was developed to investigate possible links between driver Locus of Control and risky or unsafe driving behavior.

One specific body of research conducted by John Jones and Lisa Wuebker is especially relevant to our discussion.[43] They co-developed the **Safety Locus of Control** scale. Let's describe some of their research and summarize the important findings.

One hundred and forty-three persons participated in the study. All subjects anonymously completed the Safety Locus

---

[43] John W. Jones and Lisa Wuebker. *Development and Validation of the Safety Locus of Control Scale*. Perceptual and Motor Skills. Vol 61, Issue 1, pp. 151 – 161.

of Control Scale which included statements like those in the table below.

A six-point Likert-type scale (ranging from "Agree Very Much" to "Disagree Very Much") was used for the ratings.

SAFETY LOCUS OF CONTROL

Internal Control	If workers follow all the rules and regulations, they can avoid many accidents
	Most accidents and incidents can be avoided
	People can avoid getting injured if they are careful and aware of potential dangers
	Accidents and injuries occur because workers do not take enough interest in safety
Environmental & Equipment Control	Occupational accidents are mostly caused by lack of regulations and ineffective supervision
	Most injuries are caused by accidental happenings outside people's control
	Accidents are usually caused by unsafe equipment
	Accidents are usually caused by poor safety regulations
Chance and Fate	Most accidents are unavoidable
	Whether people get injured or not is a matter of fate, chance, or luck
	Workers can do very little to avoid accidents and injuries
	Avoiding accidents is a matter of luck

In addition, all subjects completed a questionnaire which required a description of their personal histories of accidents and injuries for the preceding six months.[44] Before scoring the safety scale, subjects were placed in the following five criterion groups based on a thorough analysis of their accident questionnaires.

Group	Size	Risk	Self-reported history over the past 6 months
I	6	High	Reported 3 or more accidents, with at least one accident that caused a major injury to self, an injury to another person, or property damage.
II	7	Moderate	Reported one or more near-miss accidents that could have caused serious injury to self, injury to others, or property damage.
III	60	Moderate	Typically reported two minor injuries. Injuries tended to be superficial scratches, cuts, or bruises.
IV	46	Low	Typically reported only one very minor injury. This tended to be a superficial cut or scratch.
V	24	Very Low	Reported no accidents or injuries.
VI	15	Very Low	Reported no accidents or injuries. Received extensive safety training & education (safety professionals).

Study participants were placed into one of three accident groups. Those who did not report an accident were placed in

[44] Fifteen safety professionals participated in this study and were included as the sixth criterion group based on their accident histories and level of safety training.

the *No Accident* group. Subjects who reported one or two minor accidents were put into the *Minor Accident* group. Any persons who self-reported one or more major accidents (or serious near-miss accidents) were placed in the *Major Accident* group.

Subjects were also grouped based on their Safety Locus of Control scores. The researchers considered someone as high-risk if that person scored in the bottom quartile. These subjects were assessed as being "highly external."

An analysis showed fewer than 3% of the No Accident group were highly external individuals. This compared to 15% of the Minor Accident group and 85% of the Major Accident group.

A strong relationship was established between the safety scale groupings and the accident groupings. These results are summarized in the bar chart below.

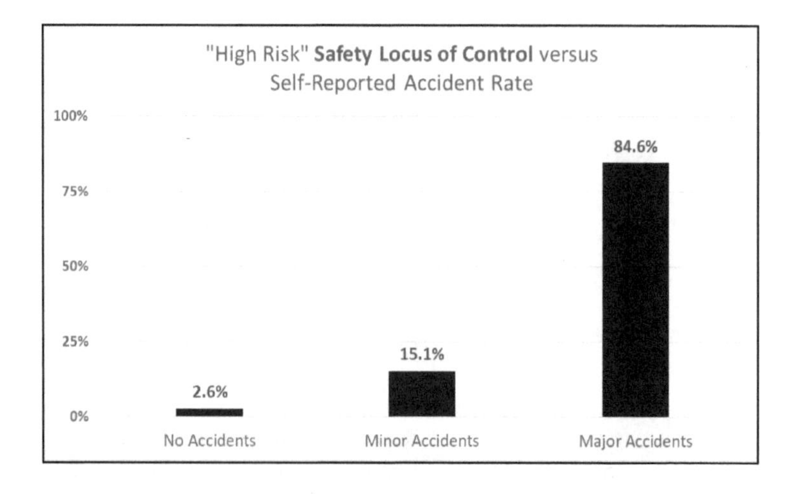

Jones and Wuebker subsequently followed their initial study with one that examined the safety attitudes of hospital workers. The Safety Locus of Control scale was assessed for the ability to predict employment applicants' propensity for on-the-job accidents. The researchers concluded hospital employees with more external safety locus of control orientations reported significantly more occupational accidents (as well as more severe and costly injuries) compared to workers with more internal safety attitudes. The results from this study further validated the Safety Locus of Control scale.

Other researchers have reached similar conclusions about the relationship between safety locus of control and safety performance. In one study, farmers who had a more internal locus of control were also assessed as being more likely to have working styles which promoted safety.

Survey Says…

In Chapter One, I outlined a few of the questions in my Safety Culture Survey. The last question on the survey is:

"How confident are you that you can work injury-free (without a recordable injury)?"

In essence, the survey participant is being asked about their *locus of control* when it comes to personal safety.[45]

About 15% of 5500 respondents to my survey indicated they were not confident in working without injury. These results indicate a surprising number of employees have an **external** locus of control when it comes to their belief in whether or not they will get hurt. In other words, 15 persons out of 100 believe if they get hurt, it will be because of fate or chance. (And the reverse is true. If they are **not** injured, they will explain their good fortune as being lucky).

[45] Some psychologists may argue this survey question is a measure of a person's *self-efficacy*. Albert Bandura defined this as the belief in one's ability to succeed in a certain situation. Locus of control and self-efficacy are very similar psychological concepts, and they are strongly related. However, we will not discuss the differences between the terms here.

Changing Beliefs

Can anything be done about employees who have a strong external locus of control for safety? Or do we simply hope they don't get hurt because of their belief system?

Fortunately, there are actions we can take to nudge these folks toward believing working safely is within their control. For example, Russ Hill describes a specific method for teaching internal control. He advocates an approach he calls the Personal Achievement Strategy™. It is built upon the hypothesis (supported by other researchers) *there is an interactive relationship between achievement and internality.*

If you can help people to set goals and experience some kind of achievement, this results in an increased belief in internal control, which in turn allows them to achieve more, which reinforces their belief in internal control, and so on. This creates an upward spiral as depicted below.[46]

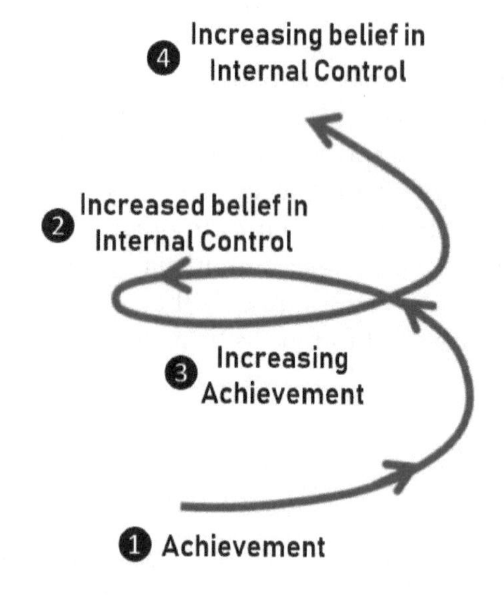

[46] Adapted from: *Teach Internal Locus of Control.* Russ Hill. Will To Power Press. 2013.

The reader who is interested in the specifics of Hill's strategy is encouraged to read his book. However, one of the cornerstones of successfully implementing this strategy is learning to set a realistic goal. The key is to set the goal so the person is challenged, but not overtaxed in striving to achieve their goal.

Connecting the dots

One of the things leaders can do to get people started on the upward spiral for safety achievement is to get them to *experience* "small wins". Sometimes we need to be reminded of how successful we have been to feel good about it and to believe we can be even better. I witnessed how one leader leveraged this idea as a means for changing the beliefs of his crew from externality towards internality in terms of safety.

Steve was an area maintenance supervisor whose crew had gone 180 days without a recordable injury. This was truly a milestone for this group. The previous longest time between recordables had been 85 days. Steve planned a little recognition ceremony for the crew before they started their day. I happened to be in the facility, and Steve invited me to attend the celebration. It was memorable.

As he circulated among the craftsmen on the shop floor, he handed out small pocket knives and thanked them for their safety efforts over the past six months. And then he said something that got everyone's attention. It was provocative.

"You know what they are saying out there in the plant?" Steve asked the group as he finished handing out the knives. "They are saying the only reason we have gone six months without an injury is we are lucky. Well, I want to know, are we just lucky?"

Silence. You could have heard a cotter pin drop.

Then one of the older guys standing in the back spoke up. "H*** no, it ain't luck!" he growled. You could see the veins in his neck stand out as he gritted his teeth.

Someone else chimed in. "Who says we are just lucky? They don't know all the things we have done to get this safety record!"

Steve got the reaction he had hoped for. "So, if it's not luck, then how did we achieve this safety milestone?" he asked no one in particular.

The discussion dam burst. One by one, different guys offered up specific actions they had taken to reduce their risk of getting hurt:

> *"We used to just blow off that pre-job checklist, but now we talk about it and check out all the potential hazards…"*
>
> *"I know we are watching out for each other more. Just the other day…"*
>
> *"No more shortcuts, that's what I see. We used to just get 'er done and didn't think about…"*
>
> *"We take a time out before a tough job and think about what could go wrong…"*
>
> *"I inspect my tools a whole lot more to make sure nothing is defective…"*

"That's right," Steve said. "We are doing all those things. I wanted you to recognize we didn't get here by luck. We got here because you guys are doing the right things to increase your safety awareness. You are watching out for one another. You are working smarter when it comes to safety. And I'm proud of all of you. But we can do better. We had some near misses. The alligators are still out there waiting to bite us. So what else can we do to keep our injury-free streak going?"

It was a simple but eloquent way to get these guys to "connect the dots" - if you do these things, you get this outcome. It was not luck or fate. It was deliberate and diligent actions. Steve was driving the spiral upward by reinforcing the relationship between effort and achievement.

- **Use Locus of Control questions as an employment screen.** There are some legal considerations to navigate before including these questions on any pre-employment test. But if you want to improve the quality of your new employees, screen out candidates who have a strong *external* safety locus of control. The research shows these potential employees are much more likely to be injured.
- **Survey employees to measure the opportunity.** The higher the percentage of employees who are not confident in working injury free, the less likely your organization will achieve its safety goals.
- **Set short-term safety goals.** Even employees with a strong external safety locus of control can be influenced to think differently if they understand the relationship between achievement and outcome. Can we go one month without injury? How about two months? Then recognize these small wins. Don't wait to go six months or a year before you celebrate, even if it is in a small way. *Winning* (working safely for a short time) promotes *more winning* (working safely for a longer time) and builds confidence.
- **Connect the dots.** People with a strong internal locus of control will naturally do this. But others may need help in making an association with specific actions and outcomes. When you are successful, have the employees tell you what has changed (à la Steve). They are more likely to make a solid connection if they do most of the talking.

PART THREE

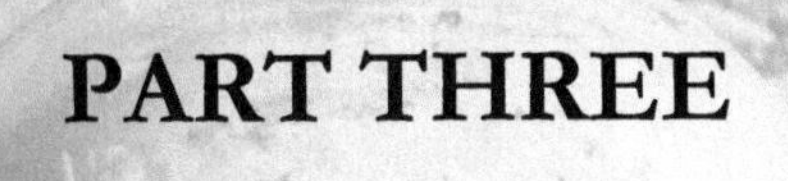

The Power of Words

"He who wants to persuade should put his trust not in the right argument but in the right word."

\- Joseph Conrad

10

MIND OVER MATTER

Imagine you are part of a team assigned to a particularly difficult maintenance job. It will take three craftsmen at least eight hours to complete this task. Your supervisor (Jeff) is coordinating a long list of planned jobs as part of a large shutdown. Before you go to the work site, Jeff provides a pre-job brief:

> *"OK guys, listen up. We have to replace the large pump in the northwest corner of the basement. As you know, it's in a very tight space with no head room and there isn't much ventilation or lighting down there, so make sure you hook up a fan and some temporary lighting. This is a critical path job, so I need you guys to get started on it ASAP. The guys on the production line will be waiting on this job before they can start back up. We're behind on shipping customer orders, so there's some heat from upper management to get in, get out, and get running. Don't take any more time than is necessary to get the pump changed out. Keep any breaks to a minimum.*
>
> *I know you guys always work at a good pace. That's why I teamed you up on this job - to get it done quickly. I'll be checking on the job every hour to see if we are on schedule. If you need anything, get me on the radio and I will rush whatever you need to the worksite. If you run into any problems and aren't sure what to do, use your judgment and do whatever takes the least amount of time. I know I can count*

on you guys to get this job done right and on time. I gotta go ... but don't hesitate to yell if you need anything!
Oh.... and be safe."

With this conversation, Jeff significantly increased the likelihood someone on this crew will take an unnecessary risk and potentially incur an injury. Why? Let's look at the pattern of words or phrases he used:

critical path, ASAP, waiting, behind, get running, time, minimum breaks, pace, quickly, schedule, rush, least, on time, don't hesitate

Do you see the pattern? What message is Jeff sending? What is the priority? What is the tone? He planted numerous seeds for the team to work quickly. In so doing, he introduced a factor which is proven to increase risk: rushing or hurrying. The team may complete the job sooner, but they will do this with a greater risk of injury.

(Notice the passing comment at the end to "be safe". That was sincere and helpful, wasn't it?)

Priming

Behavioral psychologists have proven exposing people to a series of words with the same theme can have a significant influence on their subsequent actions. This was initially demonstrated through a series of classic "priming" experiments by a number of researchers. One of them is described below.[47]

Some students were asked to come see their professor in his office. They had to walk down a long hallway to enter the

---

[47] J. A. Bargh, M. Chen and L. Burrows. (1996). *Automaticity of social behavior: Direct effects of trait construct and stereotype activation on action.* Journal of Personality and Social Psychology, 71, 230-244. [John Bargh is considered one of the pioneers in this field of study].

office where they were given a sheet of paper with a list of five-word sets. They were challenged to make a grammatical sentence from four of these words.

Here is the test:

him was worried she always

from are Florida oranges temperature

ball the throw toss silently

shoes give replace old the

he observes occasionally people watches

be will sweat lonely they

sky the seamless gray is

should now withdraw forgetful we

us bingo sing play let

sunlight makes temperature wrinkle raisins

The actual experiment was not determining if the students could arrange these words into sentences. The students had no difficulty with the test. Instead, the researchers (discretely) measured how long it took the students to walk down the hallway going TO the office and FROM the office after taking the test. Then they compared these times. The results showed it took significantly longer for people to walk down the same hallway after taking the test as compared to when they first arrived at the office.

How could this be?

The students were being primed. Words scattered throughout this test are connected with old age (worried,

Florida, wrinkled, lonely, old, bingo, wrinkle, forgetful). The researchers concluded these words triggered the "adaptive unconscious" in the brains of the students to think about the state of being old. And without even realizing it, the subjects acted old by walking more slowly!

Other priming experiments have replicated these results using other sets of behavioral word triggers.

For example, researchers primed one group by using words of aggression (*bold, rude, bother, disturb, intrude*). Another group was 'primed' with words of respect (*considerate, appreciate, patient, yield, courteous*). These individuals were observed to see how long it would take them to interrupt a staged conversation between two people. Sure enough, people who were subjected to "aggression" priming interrupted the staged conversation more than twice as quickly as those who were primed with synonyms of "respect."

Another fascinating study focused on stereotypes.[48] In a kind of "self-priming" experiment, researchers demonstrated stereotyped people themselves react differently when they are aware of the label they are given.

Many of us have heard about (or have) the stereotype that Asian-Americans are exceptionally good in science and math. Conversely, a common stereotype of females is they are not especially strong in mathematics. Both of these stereotypes are in play for Asian-American women.

The researchers divided some Asian-American women into two groups. The women in one group were interviewed and asked questions related to their race. For example, they were asked about the languages they spoke, their family's history, their ethnicity, and so forth. These women were primed with thoughts about race-related issues. The other women were asked questions related strictly to their gender. These questions addressed topics like their preferences regarding co-

---

[48] Margaret Shih, Todd Pittinsky, and Nalinin Ambady. *Stereotype Susceptibility: Identity Salience and Shifts in Quantitative Performance.* Psychological Science. 1999.

ed dorms. Afterward, all the women were asked to take the same objective math test.

The test results were consistent with the respective stereotypes. The participants who had been primed to think about being Asian-American performed significantly better on the math exam than those who had been primed to think about their gender!

These results show our behavior can be influenced by our stereotypes or our thoughts. If we think about ourselves in a certain way, we are likely to act in ways consistent with our current state of mind.

Public Pledge

How can you create an environment for self-priming? One organization found a simple way to do this.[49] They call it a *Safety Pledge.* Here is how it works: Employees are given a small printed form on a business card. They write their name on the top and keep it in their wallet. Each day before the start of the shift, the team leader or supervisor leads the crew in a recital of the items on this card. Everyone reads aloud this pledge:

1. Follow all safety rules
2. Wear all proper PPE
3. Follow STOP – CALL – WAIT for all abnormal situations
4. Never attempt to remove a stuck or jammed part by hand

This pledge is repeated every day by every person. It is a form of **self-priming**. The employees are reminding themselves and each other to adhere to a set of critical safe behaviors. And in so doing, they are putting these things foremost in their minds right before they begin working.

[49] SumiRiko in Bluffton, Ohio uses this pledge as part of their daily crew pre-shift meetings.

In fact, the items on this pledge can be considered to be *heuristics* - defined as a simple set of rules that drive behavior. The use of heuristics can be an effective means to supplement and reinforce what is important to an organization. In a safety context, here are some heuristics I have encountered:

- We care about one another
- No shortcuts
- Take two
- I've got your six
- If we need help, we ask
- Stop when unsure
- Do the right thing when nobody is watching
- Don't look the other way
- Hurrying hurts

Pre-suasion

Psychologist Robert Cialdini introduced the notion of *pre-suasion.* He defined this as the process of arranging for recipients to be receptive to a message before they encounter it. In his seminal book by the same name, Cialdini provides many examples of how channeling someone's attention leads to pre-suasion.[50] We all have a tendency to assign undue levels of importance to an idea as soon as it is drawn to our attention.

One experiment conducted by San Bolkan and Peter Andersen illustrates the subtle ways in which changing the frame of our thinking ultimately impacts our behaviors. The researchers approached people in a shopping mall with a clipboard and asked for a few minutes of time to complete a short survey. As you might expect, they did not have a very high response rate. Only 29 percent of those who were approached agreed to complete the survey.

---

[50] *Pre-Suasion: A Revolutionary Way to Influence and Persuade.* Robert Cialdini. Simon and Schuster. New York, NY. 2016.

Then they changed strategies. They introduced a simple pre-suasive opening question when they approached people by asking, "Do you consider yourself to be a helpful person?" Upon thinking about this question, almost everyone responded, "Yes." Now the subjects were asked whether they would complete the survey; the response rate skyrocketed to 77 percent!

What was happening here? By answering the pre-suasive question the subjects in the second group confirmed privately and affirmed publicly they were indeed "helpful." And what do helpful people do? They assist someone who has a few survey questions.

How could we leverage this concept to gain support? Imagine you would like some of your more experienced workers to take on the responsibility of mentoring new employees in their area. Some of these senior workers may gladly accept the challenge. Others may not. What if you pre-suasively approached these individuals and asked the following question first, "Do you consider yourself to be a team player?" Then follow this with the real question you would like answered, "Would you watch out for the new employees in your area while they go through training?"

By taking this approach, you significantly increase the likelihood of getting an affirmative response.

Let's return to our supervisor, Jeff. We can see even more clearly he was priming his employees to work at a fast pace - and thus triggering their adaptive unconscious to think about working quickly. This means they will be more likely to take shortcuts, not wait for someone to help them, skip a step in a procedure, etc. By simply hearing Jeff's words, his crew was highly influenced to take unnecessary risks. Perhaps he did not intend this to be the outcome. Nevertheless, his choice of words set the stage for risk-taking.

Positive Priming

Good news! We can use our knowledge of priming to influence others in a positive way. Let's rewind the clock and imagine Jeff giving the pre-job briefing. This time he has a very different conversation.

> *"OK guys, listen up. We have to replace the large pump in the northwest corner of the basement. As you know, it's in a very tight space with no head room and there isn't much ventilation or lighting down there, so make sure you hook up a fan and some temporary lighting. Check out the entire area before you start the job and make sure you have everything you need to do this job safely. We all know this is a big job. No cutting corners. Take whatever time you need to do the job right. It's gonna be hot down there, so you need to take a break at least every hour. Take a cooler of drinks with you to stay hydrated.*
>
> *Work at a pace that makes sense. No heroes. You guys need to watch out for each other. If you see something that doesn't look right, stop. Call me or send someone to get me, and we will talk it over. I'll be checking on the job frequently to see how you guys are feeling and if you need any relief. If you run into any problems and aren't sure what to do, take a time out to think about it. I know I can count on you guys to get this job done safely. Before I go, let me remind you to: stop when you are unsure, stay hydrated, take frequent breaks, help each other, and watch out for pinch points.*
>
> *What questions or concerns do you have?*

In this version, Jeff changed the message entirely. He was priming the crew to work deliberately and with increased situational awareness. And it was delivered in a caring tone. Let's look at his choice of key words or phrases:

check out, make sure, safely, no cutting corners, take time, take breaks, stay hydrated, no heroes, watch out, stop when unsure, check frequently, feeling, relief, help each other, pinch points, concerns

There are a numerous reasons why people take risks. Perhaps this crew would complete the job without incident, regardless of what they heard from their supervisor in a pre-job brief. However, employees who are primed with words which encourage risk-taking will most likely have a higher incident rate than those who are primed for risk awareness.

Unintentional Priming

You may be wondering whether priming or pre-suasion can actually influence peoples' decisions or behaviors beyond the hypothetical situations or experiments cited here. Let me assure you these phenomena are real. The following scenario is based on an actual event.

A large paper mill in the Midwest had gone a long period of time (nearly 15 months) without a recordable incident. Then someone was injured. The incident investigation revealed a maintenance employee had crushed his finger while removing a large motor on a maintenance shutdown. Importantly, this particular job was on the critical path for the planned maintenance schedule. The investigators interviewed the three persons involved in the incident independently.

The supervisor recalled he had stopped by to check on the job. (This was about 7.5 hours into a planned 8-hour maintenance day). He asked the two craftsmen who were assigned this task how much longer it would be before they finished. They responded that the job was nearly completed. The supervisor thanked them and left. Ten minutes later, one of the craftsmen had his finger crushed between the old motor they were removing and the frame of the machine.

When investigators talked to the craftsmen (independently), they heard a different version of the story. They each reported the supervisor told them to "pick up the pace" to get the job wrapped up.

What happened? The supervisor had unintentionally "primed" the craftsmen to hurry up simply by asking the question, "How much longer do you think you will be?" As a result of what the men *thought they had heard*, they were in a rushed state of mind. One of them was placed in the line of fire while removing the old motor, and he suffered the crushed finger.

In his defense the supervisor testified he never used those words ("pick up the pace"). Yet, this was what the other two men *heard*.

There are no villains in this story. It is simply a real-life example of how our brains work when they are primed. Because the supervisor asked the question about "time", the behaviors of the craftsmen interpreted this to mean they needed to hurry to get the job done on time. And that means picking up the pace.

To their credit, this organization implemented a policy change. It is stated something like this: On any **critical path job** you are monitoring, never lead with a question like, "How much longer will it be?" You can certainly ask this question. But it should be the third or fourth thing you say to the employees on the job – never first. Why? Once you ask this question, the people doing the work have been primed to rush or to feel time pressure, and are therefore susceptible to human error that could lead to an injury.

In addition, the site instituted a second policy change: only <u>one person</u> is permitted to ask the "how much time?" question for any critical path job. After all, when you have multiple people stopping by every 30 minutes checking on the progress of a job, you are (unintentionally) priming employees to rush and to potentially take shortcuts.

Conclusion

What is the lesson? Choose your words thoughtfully when you are in a position of influence. They carry more weight than you realize.

- **Openly express caring.** It isn't just the words you say. It's the tone of those words. People will quickly detect if you are insincere. If you genuinely care about the well-being of your employees, let them know!
- **Use positive priming.** It takes no more effort to deliver a pre-job brief where you positively prime others than if you simply go through the motions. It *does* take a little more thought. Choose words associated with the behaviors you expect. Do you really want safety to be top-of-mind? Do you want your employees to have greater situational awareness? To watch out for one another? Then make it clear by repeatedly priming your audience with these messages. Positive priming cannot be overdone.
- **Craft and communicate key heuristics.** Do your employees know what is most important? Do they know how to behave when there is uncertainty? A simple set of catch phrases can serve as guidelines or reminders of what to do in these situations. The more specific these phrases are, the more likely they will be memorable and actionable.
- **Beware of unintentional priming.** Sometimes what another person hears is not the same as what we intend to say. On critical communications it is worth reviewing in advance what will be said. Ask someone to listen to your message and provide an unbiased opinion on what they hear as the key points. Consider changing your words or phrases to deliver a clear message that matches your intent.

SAFETY LEADER'S TOOLBOX™

- **Use actions to strengthen words.** All things being equal, what you do carries much more meaning than what you say. You can use priming messages that implore people to "work safely" all day long, but if you don't back this up by truly caring about others and helping them to be successful, your words will ring hollow. (*see Part Five – The Power of Actions – beginning on page 185*).

"The best and most beautiful things in the world cannot be seen or even touched. They must be felt with the heart"

— Helen Keller

11

WHEN FEELINGS GO VIRAL

Are you aware of the extent your emotions may affect others? Your verbal and face-to-face cues are surprisingly influential in determining the feelings of those who are listening and observing. This phenomenon is known as *emotional contagion*. It is recognized by behavioral psychologists as a kind of interpersonal influence.

Studies reveal emotions (either positive or negative) spread among group members like viruses. Emotional contagion often occurs at a subconscious level. In other words, people are unaware their emotions may have been affected by another person's mood with whom they are in close contact.

But it goes even deeper than a transfer of emotions. Research by Sigal Barsade[51] has demonstrated when emotional contagion takes place, the judgment and quality of group decisions are also impacted. Let's review a study which shows how the emotions of a single person can significantly impact an entire group's performance.

Business school students were divided into small groups for a simulated management exercise. Each had to role-play a department head advocating for an employee to receive a

---

[51] *The Ripple Effect: Emotional Contagion and Its Influence on Group Behavior.* Sigale Barsade. Administrative Science Quarterly. Vol 47, No. 4. December 2002. pp. 644 - 675.

merit-based increase. At the same time, all the students were part of a "salary committee" negotiating how best to allocate a limited amount of bonus money. In essence, they had to balance getting the most for their own candidate, while maximizing the overall benefit to the company. Each group was also seeded with a confederate (an actor) who was trained to convey one of four different mood conditions:

- cheerful enthusiasm
- serene warmth
- hostile irritability
- depressed sluggishness

The researchers were able to identify several effects of emotional contagion. Groups in which the confederate had "spread" positive emotion experienced an increase in positive mood. But the emotional contagion was not limited to a spread of feelings. These groups also displayed more cooperation, less interpersonal conflict, and believed they had performed better on their task than groups in which negative emotions were spread by the confederate. In addition, groups in which people felt positive emotions made decisions that allocated the available bonus money more equitably.

When the participants were asked why they allocated the funds the way they did, and why they thought their group performed the way it did, they pointed to factors such as their ability to negotiate or the attributes of the "candidates" they had been assigned. They were completely blind to the fact their behavior and decisions (and those of their group) had been influenced by the displayed emotion of the confederate.

This study underscores the importance for leaders to be aware of emotional contagion among members of work groups or teams. Anyone has the potential to create an environment where emotions will spread. All this person has to do is visibly and consistently display their feelings and other members of the team become susceptible to catching this emotional virus. The overall effect would be positive if the emotion that spreads is conducive to a high-performing team. But it works both ways.

Perhaps you have heard the expression, "one bad apple can spoil the whole basket." Negative emotions frequently expressed by a single disgruntled employee can certainly impact an entire group's attitude and therefore their performance. These employees need to be identified and actions taken to mitigate the poisonous atmosphere from their overt negativity. This could mean a private coaching or counseling session. Or you may have to remove this individual from the team if this behavior persists.

As a leader, you can leverage this knowledge of mood contagion to create the right environment. It starts with self-awareness of your own emotions. The majority of communication occurs with non-verbal cues. These include all types of communication that do not have a direct verbal translation. Examples are body movements, body orientation, nuances of the voice, facial expressions, details of dress, and choice of objects. Time and space can also be perceived as nonverbal cues. In other words, nonverbal cues include all the ways you present and express yourself apart from the actual words you speak.

Imagine you encounter a co-worker who is usually friendly and talkative. Not only does this person fail to acknowledge you with a casual 'hello', but they walk by with their head down, tight lips, and with both hands in their pockets. When you approach them and ask if they are okay, their response is a terse, "I'm fine." Do you believe what they say or what you see in their non-verbal cues?

When a person sends a mismatched message (where non-verbal and verbal messages are incongruent), recipients almost always believe the predominant non-verbal message over the verbal one. According to Darlene Price, a body language expert, *how* we say something is more impactful than *what* we say. In some studies, nonverbal communication has been shown to carry up to 93% more impact than the actual words spoken, especially when the message involves emotional meaning and attitudes.

Before you talk to your team about any significant event or business decision, consider what emotions you would like to spread. Then deliver your message with emotions (and non-verbal cues) that match your desired outcome.

I can distinctly recall a situation earlier in my career where an executive spoke at a town hall meeting to a large group of employees. He was communicating the company was "re-focusing on safety." He wanted everyone to know the senior leaders were committed to creating a safe work environment and asked for all employees to share in this commitment.

While the message itself was a good one, it was delivered like an evening television news program. It was a presentation on power point slides with lots of charts, bullet points, and slogans. He read each slide like it was simply a collection of facts.

Everyone heard his dispassionate delivery and the empty rhetoric. You could look around the room and see very few people were connecting with this leader.

The *coup de gras* came at the end of the presentation. "Does anyone have any questions?" the leader prompted, clearly hoping there were no takers. He got his wish. Silence. We left the room. The power point slides faded into some obscure file. Nothing changed.

Do you want people to feel good about an accomplishment? Do you want them to be inspired to reach the next goal? Make eye contact. Keep your arms open while you talk. Use an upbeat tone. Don't just say congratulatory words, let them hear *and see* your enthusiasm and gratitude!

One situation which requires thoughtful use of cues to convey the right emotion is when a supervisor or manager conducts a personal safety conversation. Specifically, if the employee does not perceive from your body language, verbal tone, and actions that you are sincere, it does not matter what words you use to assure the person you care about their safety.

As Ralph Waldo Emerson said, "What you do speaks so loudly I cannot hear what you say."

Conclusion

We communicate not only with words, but also with other verbal and non-verbal messages. The emotions we express can influence how others feel. This cycle of emotions can be contagious and ultimately affect an entire group. To have the largest impact on the feelings (and actions) of others, consider the impact of emotional contagion.

How you deliver a message can have a greater impact than the message itself.

- **Be aware of your emotions.** The first step in controlling emotional contagion is to have self-awareness of how your emotions are seen by others. Ask a confidant to give you feedback in key situations on how your emotions were perceived. Use this feedback to hone your emotional control so you can learn to match your emotional state to your message.
- **Offer to provide feedback or be a mentor.** It is almost impossible for anyone to objectively assess the message they are sending with their non-verbal cues. Offer to assist your employees or coworkers in this area by providing feedback on this aspect of their communication.
- **Take a time out when needed.** It's difficult to make objective decisions when you are in an emotional state. By stepping back and engaging someone at a later time, you are less likely to allow your emotions to cloud your judgment.
- **Learn to use non-verbal cues.** Remember nonverbal communication is often more impactful than the words we say. When there is a mismatch between the two, non-verbal cues determine the message sent. When you tell someone you care about their personal safety, making direct eye contact or perhaps briefly placing a hand on their shoulder will cement the sincerity of your words.

"No one who achieves success does so without acknowledging the help of others. The wise and confident acknowledge this help with gratitude."

– Alfred North Whitehead

12

THE POWER OF ACKNOWLEDGING

Most people understand providing positive reinforcement is a proven way to encourage a desired behavior. But perhaps we don't fully appreciate how powerful the simple act of acknowledging someone's effort impacts their willingness to work and their productivity. A fascinating study sheds light on the connection between acknowledgment and intrinsic motivation.

Researchers conducted an experiment to determine if simply acknowledging a person's effort could increase their motivation to perform more work.[52] The results may cause you to reconsider how you interact with others for whom you provide leadership or direction.

The experiment was set up as follows:

A stack of papers was created where letters of the alphabet were placed in random order on each sheet of paper. Participants were given a single sheet and instructed to find all the pairs of identical letters next to each other.

When the first paper was completed, they were paid 55 cents. The participant was then asked if they wanted to complete the same assignment (finding adjacent pairs of

---

[52] *Payoff – The Hidden Logic That Shapes Our Motivations.* Dan Ariely. Simon & Schuster. New York. 2016. pp. 25-27.

letters) on another sheet of paper for 5 cents less. This process continued until the participant declined to do any more work.

There were three conditions set up in this experiment. Each is described below.

In the "acknowledged" condition, each participant was asked to write their name on the top of the paper before beginning the assignment. (By writing their names on their sheets, the participants gave explicit ownership of their work). They circled all the pairs of letters they could find. Then they walked over to the experimenter and gave him the paper. The experimenter acknowledged their work by carefully examining the completed paper. He said, "Uh-huh," and placed the sheet of paper face down on a pile on his desk. He then gave the participant two choices: (1) work on another paper for 5 cents less or (2) stop and get paid for their work. If the participant requested another paper, the process continued.

In the "ignored" condition, the participants did not write their names anywhere on the paper. Also, when they handed in their completed sheet, the experimenter didn't even look at their paper. He simply placed it face down on his desk without any kind of acknowledgment. As in the first condition, he asked the participant if he or she wanted to work on another sheet for 5 cents less or stop and get paid.

The experimenters referred to the last (and most extreme) set-up as the "shredded" condition. In this case when a participant handed in a completed sheet of paper, he or she was not acknowledged. Not only was the participant not acknowledged, but the experimenter immediately inserted the sheet into a shredder next to his desk where it was destroyed! As in the other two conditions, he turned to the participant and asked if he wanted to complete another sheet for 5 cents less.

What would you predict the behavior of the participants to be after several cycles of the same condition? For instance in the "ignored" condition, doesn't it seem likely people would be tempted to cheat? After all, no one was checking their work. Why should they bother to find all the pairs of letters on the sheet? In other words, why not earn more money for less work?

What about the participants in the "shredded" condition? Wouldn't they be even more tempted to cheat? Once they realized their work would be immediately destroyed, who would ever know if the assignment was completed at all? Why not hand in a blank sheet and get paid for doing nothing?

With this reasoning you might expect participants in the "ignored" and "shredded" conditions would choose to work longer for less money.

The results were somewhat surprising.

In the "acknowledged" condition, participants stopped when the pay rate fell to around 15 cents. By stopping at this point, the participants indicated doing more work was not worth their time.

Participants in the "shredded" condition stopped working much earlier – at about 29 cents. These results show when we are acknowledged for our work, we are willing to work harder for less pay. Conversely, when we are NOT acknowledged, we lose our motivation earlier.

What about the "ignored" condition? At what point do you think these participants stopped working?

Participants in this condition stopped working when the payment per page was about 27.5 cents – only 1.5 cents less than the participants whose work was shredded!

The authors of the study summarized the finding this way:

"If you really want to demotivate people, 'shredding' their work is the way to go, but you can get almost all the way there simply by ignoring their efforts."

Think of a time when your work was "shredded". Perhaps you had put in a lot of effort to complete a job. Then a decision-maker said your work product was no longer necessary. "Never mind," he said. "We are going in a different direction. I need you to start working on this instead..." How did this make you feel? Most likely you were deflated, discouraged, or disappointed. And definitely demotivated.

This experiment suggests if your efforts are simply ignored, it is nearly as demotivating as having your work shredded. We have all been in this situation. We diligently complete an assignment or do a job (these can be routine jobs done every day or every week). Yet no one acknowledges our efforts in even a small way. Over time, we do the minimum work required.

On the other hand, consider how acknowledging someone's efforts creates a different outcome. <u>Acknowledgment is a gift</u> from one person to another. It costs nothing, yet it can pay huge dividends. When a person is acknowledged, it conveys a message they are doing meaningful work.[53] These results show we can increase

[53] Meaningful work doesn't mean what you do has profound importance to society. It is meaningful if **you perceive** your work as contributing value to something or someone who matters. This could be your team, your family, or even yourself.

motivation simply by acknowledging the efforts of those working with us.

Dan Pink highlights this same connection between acknowledgment and motivation in his book, *Drive*.[54] Pink's review of the literature reveals there are three factors that lead to better performance and personal satisfaction: autonomy, mastery, and purpose. Autonomy is our desire to be self-directed. Mastery is our need to get better at doing things. Purpose is the notion of having a compelling reason for doing anything.

Another way to think of Purpose is this: Is what I am doing worth my effort? Is it meaningful (to someone)? Pink reminds us having a "purpose motive" is one of the factors which determines whether we are motivated to perform work. When someone acknowledges our work, we get a strong signal we are doing meaningful work. Thus we are motivated to do more of this work. That's what happened in the experiment described earlier.

Acknowledging Safe Behaviors

This simple yet powerful idea of thanking or acknowledging someone for doing something correctly or performing a task in a safe manner is essential to building a culture of commitment as defined in Chapter 1. Recall a key attribute of this culture is most safety conversations are proactive (versus reactive) in nature. Supervisors or Team Leaders engage in these proactive conversations to encourage learning and improvement as well as to reinforce desired safety behaviors. Examples of these behaviors include: not taking unnecessary risks, wearing PPE (personal protective equipment), following the safety rules and procedures, watching out for one another, and so forth.

---

[54] Drive. *The Surprising Truth About What Motivates Us.* Dan Pink. Riverhead Books. 2011.

I created a **Safety Conversation Guide** application which allows clients to document, measure and monitor personal safety conversations.[55] When this program is initially configured with the client, I strongly advise including "Thanks/Acknowledge" as one of the kinds of proactive conversations they have with employees. This is crucial because it is human nature to look for what is *wrong*, not what is *right*. This tendency is a type of confirmation bias discussed earlier.[56]

We need to proactively look for what people are doing well. Then we should thank or acknowledge the employees for doing so. It only takes a minute or two, but it ultimately can make the difference in whether these safe work habits are sustained - even when no one is watching.

Conversely, what happens if we ignore an employee who is doing something correctly – especially if it takes more effort to do it in a safe manner? For example, what if an employee takes the time to don all their required protective gear before performing a certain task. Imagine the reaction of this employee if his or her supervisor repeatedly walks by but says nothing to the individual about wearing the right PPE. At a minimum, this person is not receiving any external motivation to take the extra time to put on the protective clothing. Without positive feedback, the employee might presume his effort has little purpose other than to satisfy some arcane rule or safety requirement. They need to know their effort **has a purpose** and it is **appreciated!**

What do we mean by acknowledging someone's effort? Often it is accomplished through simple, sincere statements.

I appreciate...
Thanks for your efforts to...
This is exactly the way the procedure...
I know it took extra time for you to...

---

[55] p. 230
[56] pp. 87-95

Conclusion

When a person is acknowledged, it conveys a message they are doing meaningful work. We can increase motivation simply by acknowledging the efforts of those working with us. Proactive personal safety conversations are an effective way to communicate and positively reinforce safe work behaviors.

Are you leveraging the power of acknowledging or paying a price for ignoring?

- **Add "Thank You/Acknowledge" to your proactive conversations.** This type of interaction is just as important as other kinds of proactive conversations (e.g. toolbox talks, crew meetings, pre-job briefs, and so on).
- **Look for things being done "right."** This should be a purposeful mission. It is a planned walk through the workplace seeking out examples of safe work behaviors or habits. Once you observe something, stop and tell the person, *"Thank You"*. Be specific in your feedback so that person knows WHAT he is being praised for and WHY it is important he is doing it.
- **Take time for the unexpected positives.** Be mindful of the small things people are doing in a safe way and point these out every day. For example, if you are walking to a meeting and see an employee who is exhibiting a desired safe behavior, take one minute to acknowledge this. Thank them.
- **Set an R^+ goal of 5 to 1.** While there is no "set" ratio, several studies suggest individual or team performance is highest when there are *about* 5 times as many positive reinforcements (R^+) as negative reinforcements (R^-). Keep this in mind as you provide feedback.

"If you expect nothing from anybody, you're never disappointed."

— Sylvia Plath, The Bell Jar

13

YOU CAN DO IT!

Earlier in my career, I had the opportunity to work with a group of four supervisors at a manufacturing site. I spent a week with each supervisor, with the objective of getting to know each person better. After a month rotating among the supervisors, it was clear there were stark differences in how each of them related to their respective crews. The contrast in styles was greatest when comparing Mitch with Harold.

Mitch considered himself to be "old-school" and was proud of it. He had spent nearly twenty years in various line positions at the plant and eventually worked his way into a senior operator role before being promoted to supervisor. He was a no-nonsense guy who ruled with an iron fist and a commanding voice. His philosophy was to set the rules and hold people accountable when they were violated. Mitch believed his primary responsibilities were to "keep the line running" and to "make sure no one does anything stupid." During the shift, he could often be found in the supervisor's office area, unless the line was down for some reason. His crew tended to have the least senior people mainly because there was a lot of bidding to move to another supervisor's crew.

Harold also spent many years as an operator in the same facility before accepting a supervisor position. He had a calm demeanor and spent most of his time on the floor listening to his crew members. He frequently answered any questions with a question of his own, "What do you think we should do?" Harold challenged his crew to come up with solutions, not just to identify the problems. I would overhear him privately praising each person, telling them they were among the best operators he had ever been around. When someone made a mistake, he would make it a point to ask the individual what lesson was learned and what we could do differently the next time. Harold's crew had the most senior people. It was clear they respected Harold and valued the opportunity to work on his crew.

It is no surprise there were significant performance differences among the crews. Harold's crew had superior results in productivity, yield, safety, and absenteeism. In comparison, Mitch's crew had the lowest output, were more likely to have an injury, and used all of their available "sick days" throughout the year. While there are many aspects of Harold's leadership that contributed to his crew's performance, the one I want to focus on here is setting expectations.

Although I had heard about the notion of people living up to expectations, this was a real-life case study on how expectations can either have a positive influence (à la Harold) or a powerful suppressive effect (à la Mitch). Social psychologists refer to this phenomenon as the Pygmalion effect.[57]

The Pygmalion (or Rosenthal) Effect occurs when **people perform to the level of expectation placed upon them**. The effect is named after a Greek myth. (Pygmalion was a sculptor who fell in love with a statue he had carved. He wished for a bride as lovely as his statue. The ivory girl he had sculpted subsequently came to life).

This effect has been confirmed in various settings. In sports many athletes and teams attribute their success to

---

[57] Eden, Dov. *Leadership and Expectations: Pygmalion Effects and Other Self-Fulfilling Prophecies in Organizations.* Leadership Quarterly, 3(4), 271-305. (1992). JAI Press Inc.

high expectations from a coach or them-selves. In leadership applications, it is a type of self-fulfilling prophecy (SFP) in which raising manager expectations boosts subordinate performance. Managers who expect more of their employees lead them to greater achievement. SFP is the process through which the expectation an event will occur increases its likelihood of occurrence. Expecting something to happen, we act in ways that make it more likely to occur.

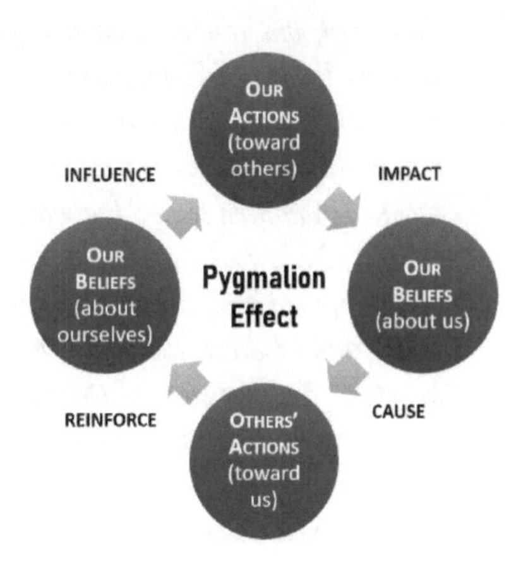

As a leader it is easy to forget most people are willing to step up to a challenge. We just need to set the expectation, provide them with the opportunity, and ask how we can help them to be successful.

Let me share a personal experience about safety expectations which occurred many years ago.

One of my summer jobs during college was working at a busy warehouse. My main task was to fill wooden pallets with various orders of canned fruit or juice products. Forklifts loaded the pallets on a trailer for shipping.

I remember the day I filled out the employment application. The job was on second shift. It was hard to find anyone who wanted to work these hours.

The woman from human resources asked me if I could start working the same night. I showed up 30 minutes before my shift for orientation. While I don't recall everything that was said, the supervisor's safety expectations were memorable. The speech from Lyle went something like this:

> *"Most of these guys have been working here for more than 15 years. So ask them anything you want to know. The work isn't*

that hard, but you can expect to get a few minor injuries before the summer is over. Nothing serious - maybe a gash from a box cutter or a sore toe from a case that is dropped accidentally. Oh, that reminds me: no open-toed shoes. There's a first aid kit in the break room. If you need something more than a bandage or ointment, come see me. Now look, the number one thing you need to remember is those guys running the forklifts are moving fast. The sooner we get these trucks loaded, the more time we all have at the end of the shift to relax. So stay clear of them at all times. They have the right-of-way in the aisles. Any questions?"

What questions would an 18-year-old ask? I had none.

I shouldn't have been surprised when I noticed the guy who was training me had a scar above his left eyebrow. When I asked him about it several weeks later, Rick said a carton started to slide off the top of a second-tier pallet. When he reached up to push it back, it fell off and caught him just above his eye. He shrugged, "It just wasn't my day. At least it didn't fall on my foot - now that would have really hurt!"

Looking back at this experience, it is obvious safety considerations at this facility were almost nonexistent. From my perspective, the most glaring factor in the casual attitude towards safety was the expectation of the supervisor (and his manager). He expected his employees to be injured. Not seriously, of course. But he expected them to get a few nicks and bruises along the way. It was just part of the job.

We get what we expect

In retrospect, I was an eyewitness to the Rosenthal Effect which states: The greater the expectation of achievement, the greater the level of success. Unfortunately my situation demonstrated the reverse: The lower the expectation of achievement, the lower the level of success.

Rosenthal and Jacobson conducted a classic study that demonstrated how our expectations can influence people.[58]

The study was set up like this: To test whether a teacher's expectation of student performance affected student achievement outcomes, researchers gave an IQ exam to elementary school students. The students were ranked based on scores. Teachers were told the top 20% of students had high potential to succeed. Further, the teachers were provided with the names of the students in the top 20%.

Here is the interesting part. What the teachers didn't know was they had actually been given a *random list of names*. The researchers came back at the end of the school year and administered the exam again to the same group of elementary students.

What they found was astonishing: The second-graders and third-graders who had been labeled as "high potential" at the beginning of the year had advanced significantly beyond their peers as measured by higher IQ scores. But recall these students had been randomly assigned to this list!

The researchers concluded the teachers' expectations of student achievement actually became self-fulfilling. Those students labeled "smart" actually became so. The teachers (consciously or unconsciously) paid closer attention to these students or treated them differently when they were having difficulty. The teachers and students at this elementary school believed in the existence of "smart" and "not-so-smart" students. They made it their reality.

Expect to work safely

Lyle expected the guys on his shift to have a few minor injuries from time-to-time. And so it was. Further, this belief was held by the guys who worked for Lyle *and they accepted it!*

---

58 Rosenthal, R., & Jacobson, L. (1968). *Pygmalion in the classroom: Teacher expectation and pupils' intellectual development.* New York: Holt, Reinhart & Winston.

As a supervisor or manager, what expectations do you have for others? How do you share these expectations with your employees? It isn't just what you say, but what you **do** that gives you credibility. As the old adage goes, "Your actions are so loud I can't hear a word you are saying."

I have no doubt if Lyle had set a different level of safety expectation for his employees, their safety performance would be dramatically better. Why? Because he would be (consciously and unconsciously) providing a focus on safe working habits for everyone. He would be coaching his crew on situational awareness and influencing them to watch out for one another. Lyle and his crew would believe in the existence of "safe" employees. And it would become their reality.

Thankfully I don't recall any injuries while working that summer. I never went back to that job. But I vividly remember Lyle's distinct limp. One of my co-workers told me Lyle was in an accident many years ago. The forklift he was operating tipped over when he rounded a corner at excessive speed and tried to avoid a pedestrian.

I can't help but wonder what safety expectations Lyle's supervisor had...

Conclusion

People tend to perform to the level of expectation placed upon them. In the workplace those with authority or influence have a significant role in determining how employees will perform simply by the expectations that are set. If your supervisor or manager believes you can meet a specific challenge (and shares their confidence in your abilities), the chances of your success are much greater than if they openly express doubts about what you are capable of achieving.

What expectations are you setting for your employees?

- **Perform a personal expectations assessment.** Do you believe you can lead (part of) an organization toward an injury-free workplace? If you don't believe this is possible, you have a low chance of success. If you think it can be accomplished, you will be inclined to align your behaviors and actions to achieve that goal.
- **Don't label employees as either "safe" or "not-so-safe".** There are a number reasons why some people seem to be more prone to injury than others *(see page 107 for one of these)*. Avoid contributing to the propensity for certain people to have accidents by assuming they will get hurt more often. If you expect they will be injured, this is more likely to be the outcome because you may (unconsciously) act in ways that support your own beliefs. Instead, work with them to change their perspective. Help them to see the connection between effort and outcome when it comes to safety.
- **Set an expectation everyone can work injury-free.** Make it clear there are no exceptions. Then work with these individuals to change their own expectations about being injured. They may need more coaching in some areas than their peers to be successful. Collaborate on developing personal safety action plans that enable everyone (including these so-called 'injury-prone' employees) to be part of an injury-free workplace.

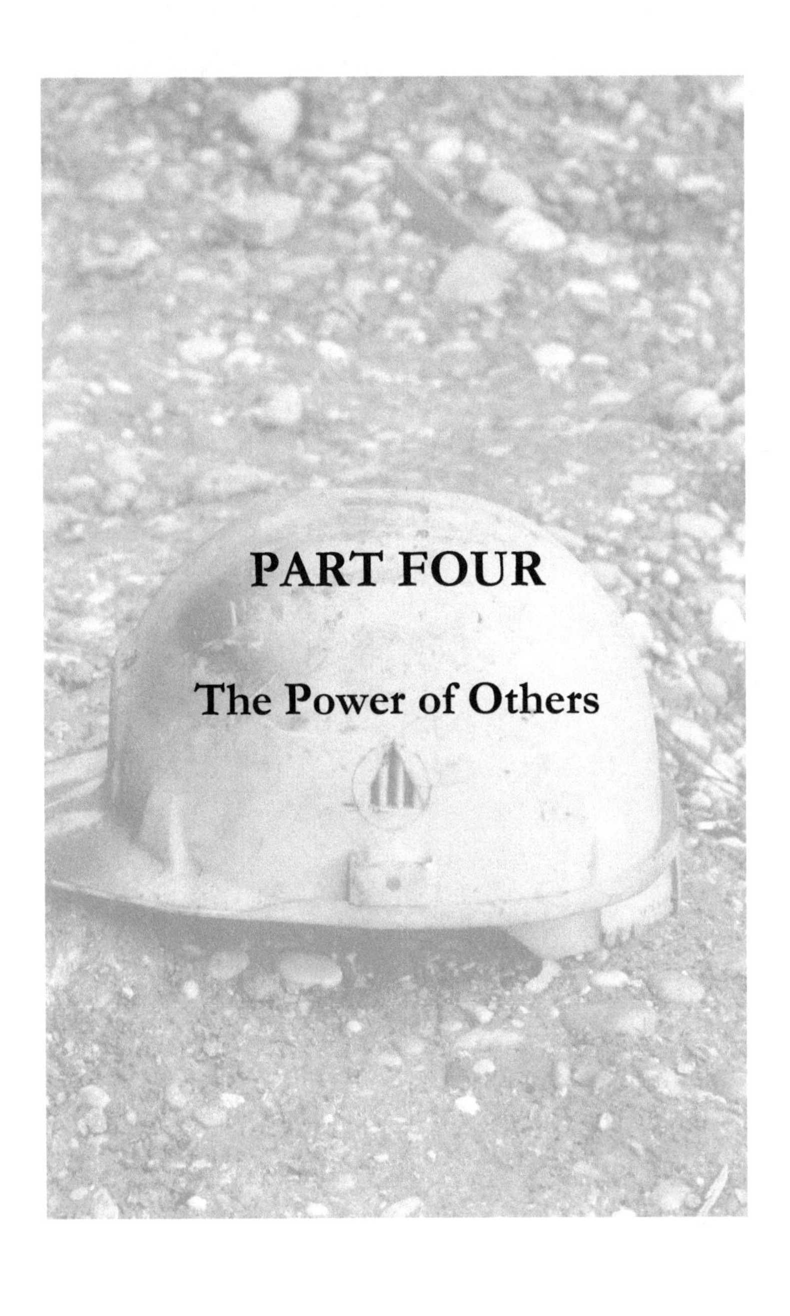

PART FOUR

The Power of Others

"Sometimes if you want to change a man's mind,
you have to change the mind of the man next to him first."

— Megan Whalen Turner

14

RESPECTED AND CONNECTED

Sustainable culture change is difficult. A common view is the work required to shift the mindset of any large organization is the responsibility of leadership. Certainly the senior leaders of the organization are accountable for setting the vision and supporting it by what they say and do.

Unfortunately, many leaders approach this challenge by delivering a message via well-written power point slides to the masses. The assumption is all they need to do is explain what the company is about and how everyone should be aligned to be successful. Perhaps for a short while this strategy results in an uptick in the "hoped-for" behaviors.

But it does not last. As time passes most of the employees slip back into their comfort zone. They see no reason to change anything. "After all," they think, "This too shall pass. If I wait long enough, someone else will come along with a different flavor-of-the month. Maybe that one will taste better."

There are many potential reasons leaders may be unable to get an entire organization aligned to accept a new way of thinking and acting. In this chapter I will focus on a single consideration.

We accept change at different rates

Everett Rogers published his theory on the *Diffusion of Innovations* in 1962.[59] It is a theory that seeks to explain how, why, and at what rate new ideas and technology spread through cultures. The book (now in its fifth edition) says diffusion is the process by which an innovation is communicated through certain channels over time among the members of a social system. The innovation or idea must be widely adopted in order to self-sustain.

In his book, Dr. Rogers tells a fascinating story of how he was prompted very early in his career to study how new ideas were adopted by the masses.

Immediately after graduating with a Ph.D. in sociology, Dr. Rogers accepted a job working with an agricultural extension service in Iowa. His primary responsibility was to work with the local farmers and encourage them to use newly developed varieties of corn which were proven in field tests to produce crops with higher yields, as well as being more disease-resistant. As a result, these strains of corn were more profitable than the current varieties.

Unfortunately, Dr. Rogers quickly learned he couldn't connect with the farmers. He was a college-educated young man who had never plowed a field or planted corn. All his academic knowledge didn't mean anything to the farmers. He lacked credibility.

He realized he needed to convince at least one farmer to try one of the new strains. That way, he reasoned, once this crop was proven to have higher yields, all the other farmers would follow suit and adopt the new innovation in corn seed.

After a while Rogers persuaded a farmer to try one of the new strains. There was one wrinkle, however. This farmer was unlike his neighbors in many regards. He drove a Cadillac, not a pickup truck. He wore Bermuda shorts, not jeans or overalls. Basically he was a social outcast in the farmer community.

---

[59] Rogers, Everett M. (1983). *Diffusion of Innovations.* New York: Free Press.

Nevertheless this fellow planted the corn and indeed enjoyed record yields that harvest season.

Yet no one followed his lead and adopted the new higher yield strains of corn. The reason: this renegade farmer wasn't one of them. You can almost hear the other farmers muttering, "There's no way I'm going to plant the same corn as that weird guy who calls himself a farmer. I don't care what the yield is!"

This experience is what launched Rogers' research into explaining why some ideas are adopted while others are not. He also set out to discover why certain individuals have more influence in encouraging people to accept an innovation or new idea than others. Rogers looked at research from over 500 diffusion studies across a number of fields: anthropology, sociology, technology, and education.

Rogers' theory says within the rate of adoption there is a point at which an innovation or idea reaches critical mass. The people in any social system who are exposed to the new idea can be placed in segments depending upon their willingness to adopt the new idea or accept change. Rogers named five categories of "adopters": Innovators, Early Adopters, Early Majority, Late Majority, and Laggards.

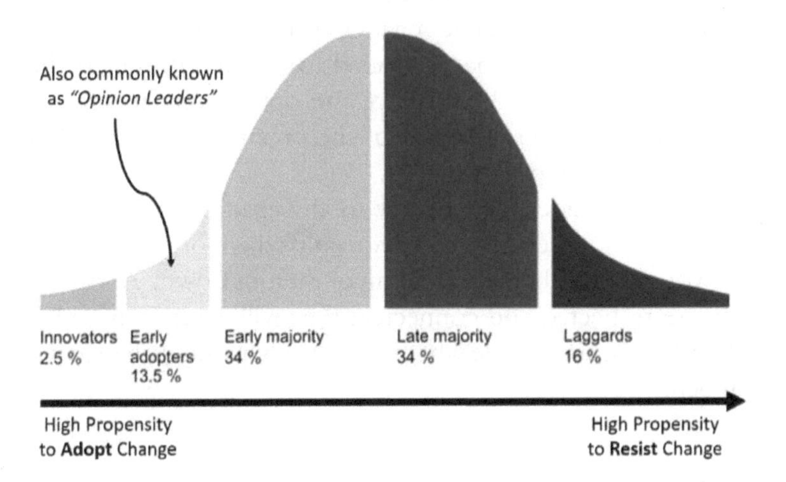

Don't try to persuade the majority

Dr. Rogers submits there is a primary group you should focus on if you want to get a new idea adopted or to make change happen. It is not the Early Majority or Late Majority (the 2/3 of the population in the center of the adoption curve). Obviously it is not the Laggards. As the name implies, they are the last to accept anything new and will do so only as a final option.

But here is the surprising part. It is not the Innovators either. Even though these people are the first to try something new, they tend to be social outliers who generally do not have strong influence on their peers. (The farmer in the Bermuda shorts fits into this group). Plus true Innovators only make up 2% - 3% of the organization.

Rogers claims the real power brokers are the Early Adopters (*Opinion Leaders*). This group (about 13.5% of the population) is open to new ideas. But the key attribute that makes them valuable as agents of change is they are **respected** by a large number of their peers. They are also held in high regard by many people; therefore, they are socially **connected** to the network of the organization.

It is clear sustainable change can only happen if the Opinion Leaders can be persuaded to join the cause. If this group can be enlisted, you have the opportunity to reach a tipping point[60] where the rest of the organization will follow their lead and adopt the change.

Opinion Leaders are the key to any change effort and are powerful influencers. Whether you enlist them or not, they will give your ideas either a thumbs up or thumbs down. And since they are respected and connected, they will exert their widely felt influence and decide whether (and at what rate) change will happen.

---

[60] Gladwell, Malcolm (2002). *The Tipping Point: How Little Things Can Make a Big Difference.* Little, Brown, and Company.

Identifying Opinion Leaders

Now that we know these people are critical to our plans to implement any major change, how do we identify them?

Since Opinion Leaders are employees who are among the most admired and connected to others, you probably know who they are if you have been in the organization for a number of years. You can validate your own view by simply asking people from across the organization to make a list of the employees who they believe are the most influential and respected. One question you may ask when soliciting these names is, "If you were to go to just two or three persons for advice on a problem you are having, who would you approach?" Then gather the lists and write down those who are named most frequently. These are opinion leaders.

It is important to note an opinion leader can be influential with their peers in either a positive or a negative way. In addition, don't presume these people have some kind of formal authority. Many are informal leaders with no direct authority.

Opinion Leaders can tip the scales

I observed first-hand how this critical group of employees were difference makers.

After a number of near miss events involving pedestrians and forklifts, a large warehouse and shipping facility decided to implement an across-the-board policy which required everyone to wear high visibility vests everywhere except the office areas. The leadership communicated this policy to all the employees in a series of meetings. The general response was not positive. Quite a few people grumbled about the new safety vest requirement and cited numerous (weak) excuses why some employees should be exempt from this new policy.

The facility manager was taken aback by what he heard in these meetings. The policy was not negotiable. It was clearly the right thing to do. He had hoped more people would have accepted this new policy strictly on its merits of reducing risk. He decided to enlist some help to enable this change.

The leadership team had a fairly good idea who the Opinion Leaders were within the facility. They cross-checked their own lists with the supervisors on each shift to make sure they did not miss anyone. The Opinion Leaders were invited to a meeting where the management team outlined the reason for the new policy. They explained that requiring everyone to wear high-visibility vests was a simple but effective way to reduce the likelihood of a forklift operator inadvertently colliding with a pedestrian because they were not visible at times when walking on the warehouse floor. The facility manager closed by asking this group, "Can you help?"

Interestingly it became obvious in this meeting there were Opinion Leaders *within* the Opinion Leaders! You could see everyone's head turn toward John before anyone spoke. John was not a union officer. He had no formal authority. He was actually soft-spoken. But he had been at the facility since it had opened over 30 years ago. And he commanded respect.

John cleared his throat and said, "I've seen my share of close calls when it comes to forklifts and people. A buddy of mine had his toes run over a number of years back, but his steel toes saved him. Why wouldn't we want to do this? Seems reasonable to me."

And with this endorsement, the others nodded their heads in agreement. The policy was put in place two weeks later, with minimal fanfare. Within a month, vests were as common as safety glasses within the facility.

Conclusion

What predicts whether an innovation / idea / change is widely accepted (or not) is whether the Opinion Leaders embrace it. Why? They are socially connected and respected. The rest of the population will not adopt the new practices until the Opinion Leaders do.

- **Identify your Opinion Leaders.** Using the method described earlier, develop a list of internal influencers. Be comprehensive in your search. Candidates can come from any part of the organization. Involve them in planning for the change. Ideally they would become advocates for the proposed future state.
- **Don't overlook or avoid those Opinion Leaders who *negatively influence* others.** They don't need to become advocates. Rather you need to find a way to have them maintain a 'neutral' posture on any major change effort or idea.
- **Populate key safety initiatives with Opinion Leaders.** Here is one example. A number of factors determine the success of safety committees and safety-related activities. These include senior-level sponsorship, a clear mission, and strong organizational or facilitation skills. Another success factor is the membership of these groups. If possible you should work to strategically nominate/assign/place one or two Opinion Leaders on the most high profile safety committees or groups. Their influence could be the difference in whether organization-wide initiatives are embraced. By participating in these groups, they are also role models for employee engagement in safety.

"If you follow the crowd, you might get lost in it"

— Anonymous

15

IF YOU WILL, I WILL

We are influenced by the actions of others more than we may care to admit. Many researchers have confirmed *social influence* has a powerful effect on our decisions.

We experience many forms of social influence, although we probably don't think about it. Perhaps you purchased something after hearing about it from a friend or family member. Or you may have joined an organization or club because someone you know is a member. Throughout our lives we have been powerfully persuaded or casually nudged thousands of times to make a decision or take an action because of social influence.

Indeed, the authors of *Influencer* contend there are six sources of influence.[61] They refer to one of these influences as social motivation (although most of us think of this as peer pressure).

Pedro Gardette of Stanford recently conducted a study that supports this concept.[62] He wanted to measure the effect of social influence on the purchasing patterns of airline passengers.

---

[61] *Influencer. The Power to Change Anything.* Patterson, Grenny, Maxfield, McMillan, & Switzler. McGraw-Hill. 2008.

[62] *Fellow airline passengers influence what you buy.* Stanford Business. February 6, 2015. http://stanford.io/1vwglQh

Gardette looked at consumers in pairs with each person sitting in the same kind of seat and flying economy on the same flight. One passenger in the pair was in the treatment group, meaning the person observed a purchase by the passenger seated next to him or her. The other person in the pair, the control, was seated in the same kind of seat one row ahead but did not observe a purchase.

The study found passengers were **30%** more likely to buy something if someone next to them made a purchase first. This did not hold true if they saw a purchase by someone sitting behind them or diagonally in front of them. It was only the person next to them that affected their buying behavior.

If two strangers could have this much influence over each other, Gardette wondered if a friend might influence another person's decision to buy. To test this theory, he analyzed the purchasing behavior of people traveling together under the same reservation number. (The assumption is these persons knew each other). He found the likelihood a passenger will buy **doubles** if the person who is next to them making a purchase is someone they know.

His conclusion: Our peers can have a significant influence over whether and what we buy.

Choosing the obvious wrong answer

Another form of social influence was demonstrated in a series of classic experiments by Solomon Asch.[63] He was interested in measuring *conformity,* defined simply as yielding to group pressures.

Using a line judgment task, Asch placed a single study participant in a room with seven confederates (who were part of the experiment).

---

[63] Asch, S. E. (1951). *Effects of group pressure upon the modification and distortion of judgment.* In H. Guetzkow (ed.) Groups, leadership and men. Pittsburgh, PA: Carnegie Press.

The confederates agreed in advance what their responses would be when presented with the visual line test. The real participant did not know this.

Here is a representation of the visual line test:

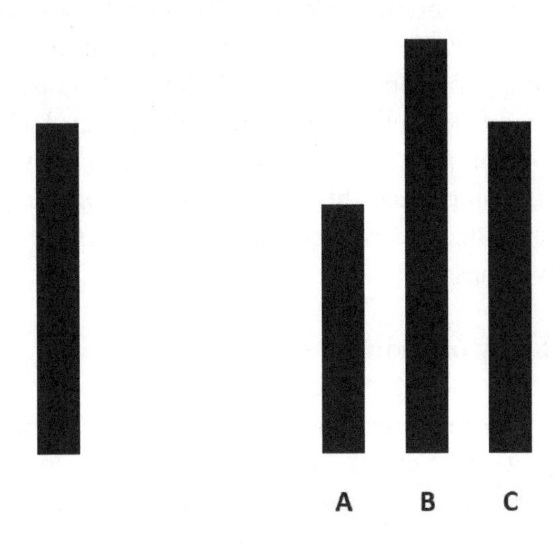

Each person had to state aloud which comparison line (A, B, or C) on the right was most like the target line on the left. The correct answer (C) was obvious. One at a time, the confederates consistently gave the *wrong* answer (choosing either A or B). The study participant was positioned in the room so he always gave his answer last.

Asch was interested to see how often the study participant would conform to the majority view. (There was also a control condition where there were no confederates, just real participants).

On average about one third (32%) of the participants who were placed in this situation *conformed* to the clearly incorrect majority in the room! (With no pressure to conform to the confederates, less than 1% of participants in the control group gave the wrong answer).

Since this original experiment decades ago, many other researchers have replicated Asch's experiment with similar findings.

Conformity can occur even in the absence of overt group pressure in a controlled setting like the Asch experiments. Robert Cialdini and his colleagues found college students were more likely to litter when they had just seen another person throw some paper on the ground.[64] They also observed these students were least likely to litter when they had just seen another person pick up and throw paper into a trash can. The researchers interpreted this as a kind of *spontaneous conformity*—a tendency to follow the behavior of others, often entirely out of our awareness.

Should I stay or should I go?

Social or peer influence goes beyond the marketing world. It has a powerful effect on our behaviors in the workplace. These influences can be the difference between working safely and taking unnecessary risks. To emphasize the impact of these influences, let's take a look at a situation everyone has experienced.

You are standing with a group of strangers on the sidewalk at an intersection with a cross walk and pedestrian signals. The signal clearly indicates *"Do Not Walk"*. There is only one car approaching the intersection, but it is several blocks away. One of the persons in this group starts to walk at a brisk pace to cross the street. What happens? Often a few people follow his lead and cross, ignoring the pedestrian signal. What would you do? Why? If you cross the street as well, you have just been socially influenced.

---

[64] Cialdini, R. B., Reno, R. R., & Kallgren, C. A. (1990). *A focus theory of normative conduct: Recycling the concept of norms to reduce littering in public places.* Journal of Personality and Social Psychology, 58, 1015–1026.

Follow the leader

It is easy to see how social influence is prevalent in the workplace. In a safety context, it can result in either desired or undesired behaviors. Many employees work in crews or teams. The culture of these groups is significantly impacted by social influences, especially those from the Opinion Leaders (Recall from the previous chapter these are individuals who are held in higher regard by their peers).

Imagine some crew members have drifted away from following the standard safe operating procedure by taking a shortcut. If a new employee observes this behavior, they immediately experience social influence. What happens the next time they are performing this same task with one of their co-workers? While we might convince ourselves WE would not take this shortcut, social research shows a significant portion of the time we would be influenced to take the same risk if others are doing it.

On the other hand, our peers can have a powerful positive influence on our behaviors. If we see someone perform a task with diligence and self-awareness, we are more likely to imitate this behavior; even more so if the person who is serving as the role model is an Opinion Leader.

Conclusion

Social influence is a powerful force. It can be a critical factor in the choices employees make about how to perform tasks. A surprising number of people will make the "wrong decision" because of the need to conform. They do not want to be seen as a social outcast.

Effective leaders understand how to use social influence in a positive way to reinforce desired behaviors and to affect change. It is unlikely most people will accept or adopt new ideas unless some kind of social influence is included in any change strategy.

SAFETY LEADER'S TOOLBOX™

- **Be aware of conformity and procedural drift.** If an entire work group has drifted from performing a task using the standard procedure, there is a good chance conformity has contributed to this behavior. With every successful completion of the task without incident, each group member is reinforced to continue the practice in part because of conformity to group pressure.
- **Couple conformity with Opinion Leaders.** Whenever possible, leverage conformity by positioning Opinion Leaders to be the trainers for many critical tasks. As we have seen earlier, their influence is powerful. We want employees to conform to safe work habits role modeled by Opinion Leaders.
- **Support respectful disagreement and questions.** You want your employees to have a questioning attitude, especially when it comes to safety. To encourage this go out of your way to provide positive feedback to anyone who raises a sincere question about a procedure or process. This will strengthen their confidence to 'stand their ground' if they see at-risk behavior. They will be less likely to succumb to group pressure.
- **Avoid group-influenced decisions.** If you want independent viewpoints, solicit these opinions in a way that protects anonymity. Some examples: (1) Take people aside and ask them privately for their views. (2) Facilitate silent brainstorming sessions in meetings rather than doing a "round-robin" exercise where everyone shares their ideas aloud.

- **Create spontaneous conformity.** What do you want employees to start doing more often when it comes to safety? Plant 'seeds' by modeling this behavior at strategic times and places. If you do these things frequently and consistently, people are likely to (subconsciously) follow the same behavior. Use your imagination to determine the kinds of safe work habits or behaviors you can easily and publicly model for others to observe.

"To say nothing is saying something."

— Germany Kent

16

SILENCE IS NOT GOLDEN

Victor has over 20 years of experience in the warehouse. You have a few years of industry experience and were just hired a few weeks ago by the company. Today you are working as a team unloading pallets of packaged materials which were delivered from the dock. As both of you approach the first pallet, Victor takes a position directly in front of the strapping that is straining under tension. You see this puts him in the line of fire. Instinctively you take a step back when Victor pulls a pair of snips from his pocket to cut the strapping...

Do you speak up? Do you stop him? Are you sure?

Perhaps you would say something. But a surprising number of people in this situation would stay silent. Their thought process would be something like, "Surely he must know how to perform this task safely. He's done it thousands of times. I'm the rookie here. Who am I to question his experience and job knowledge?"

As explained in the previous chapter, peer pressure is a powerful social influence. Most of us are fearful of being considered an outcast if we are the dissenter, especially if we have less informal authority than other people in our natural work group.

You may think of peer pressure as overt statements from co-workers. "Look, this is the way things are done around here." But this is not always the case. In the scenario above, Victor did not have to remind you about his seniority and experience. It was implied and understood.

Propensity for silence

Over 5500 employees have taken my eight-question safety culture survey. One question from this survey is designed to gauge an employee's willingness to speak up:

> *How comfortable are you in stopping and talking to a co-worker if you see them taking an unnecessary risk* ***even if they are more senior than you?***

People who answered this survey question and selected the lowest rating (on a scale from 1 to 5) described their response as "Definitely Uncomfortable. It's not my job to question someone else's work, especially if they are very experienced."

Those who chose the highest rating on the scale assessed themselves as being "Very Comfortable. I have a responsibility to speak up to anyone regardless of their authority. I would expect them to do the same for me."

The percentage of responses on the lower end of the scale (ratings of 1, 2, or 3) were combined to attain the overall percent of employees who were *not comfortable* (to some extent) in speaking up.

A comparison of responses to this question from two organizations is given below. Each organization is at a very different place along the *Safety Leadership Continuum*™. One is in the lower quartile, while the other is in the upper quartile on this scale.[65]

---

[65] See pp. 10-12 for a description of this *Continuum.*

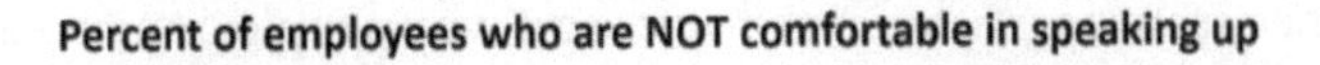

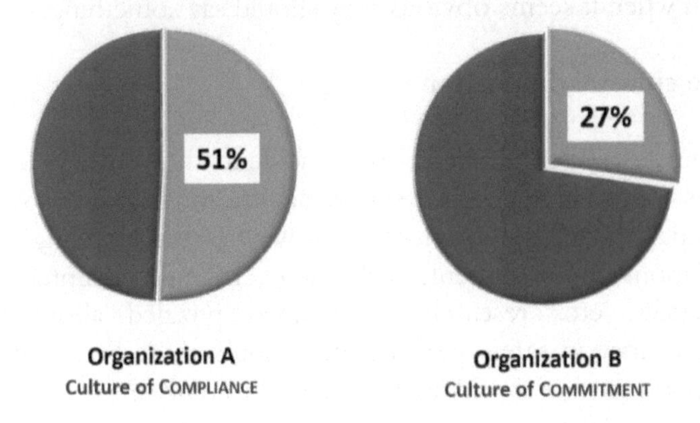

Organization **A** is driven by a culture of compliance. They have a poor safety performance relative to their industry peers. Half of the employees who completed the survey reported they were not comfortable in speaking up.

Organization **B** has a culture of commitment. They also have a good safety record. However, even in this organization more than 1 in 4 persons were not comfortable in speaking up.[66]

These findings confirm a significant number of employees hesitate to speak up when they see someone taking an unnecessary risk. If even one individual doesn't speak up when they see someone taking a risk, it could be the difference between whether or not they or their co-worker are injured. Staying silent is a workplace behavior that can have tragic consequences.

Before we can identify methods to improve the likelihood someone will speak up, we need to understand some potential factors that contribute to this behavior. A study of global

---

[66] Benchmarking data for this specific survey question indicates the median proportion of employees who are somewhat uncomfortable speaking up is 24%. This compares to only 2% of survey respondents who reported they were uncomfortable speaking up in the organization that scored highest on the *Safety Leadership Continuum*™. The complete data table is presented on p. 243.

plane crashes revealed a major reason why people remain silent even when it seems obvious they should say something.

The cultural dimension

An analysis of commercial airline incidents showed plane crash rates (per million departures) are significantly different when compared across countries. However, after ruling out variations in equipment, technology, training, maintenance practices, etc., researchers remained puzzled about this discrepancy in safety performance. Could there be a cultural component that would explain these differences?

Geert Hofstede, a Dutch psychologist, developed the **Power Distance Index** (PDI) as part of his work to understand and measure certain cultural attributes. Power Distance is concerned with attitudes toward hierarchy, specifically with how much a particular culture values and respects authority. The greater the PDI, the less likely an employee will disagree with their superior (or someone who has more power).

To determine the PDI, Hofstede conducted cultural surveys in nearly two dozen countries. He asked questions such as:

- How frequently, in your experience, does the following problem occur: employees being afraid to express disagreement with their managers or superiors?
- To what extent do the less powerful members of the organization accept and expect power is distributed unequally?
- How much are older people respected or feared?
- Are power holders entitled to special privileges?

When Hofstede plotted the PDI associated with the pilot's native country against their respective plane crash rates, he determined there was an association between these two factors. The pilots from countries with the highest Power Distance Index were over 2.5 times more likely to crash than the pilots from countries with the lowest PDI.

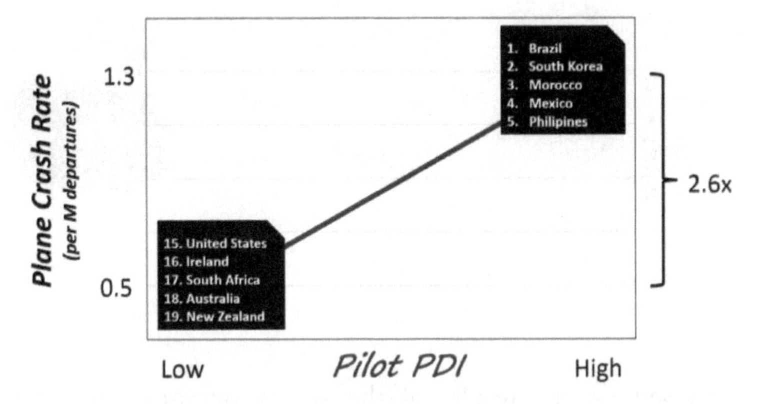

Moreover, plane crashes are much more likely to happen when the captain is in the "flying seat" even though they share the flying time about equally. Why? The evidence suggests the junior officer in the cockpit is reluctant to question the captain (who has more authority) when he or she sees something which may be an unsafe decision.

The higher the PDI of the flight crew's native country, the less likely the co-pilot will be to speak up when something does not seem right. This reticence is believed to be a significant contributor to the higher crash rate.[67]

Let's consider these implications in the context of a non-aviation workplace. Just as the cultural norms among nations are not the same, the norms among organizations can be extremely different.

Imagine if we measured the PDI of those companies who have an excellent safety record as well as the PDI of organizations with poor safety performance. My hypothesis: We would find a relationship between organization PDI and

---

[67] Geert Hofstede. *Culture's Consequences: Comparing Values, Behaviors, Institutions, and Organizations Across Nations.* Sage Publications. Thousand Oaks, CA. 2001

employee injury rate (analogous to the pilot PDI vs crash rate study described earlier).

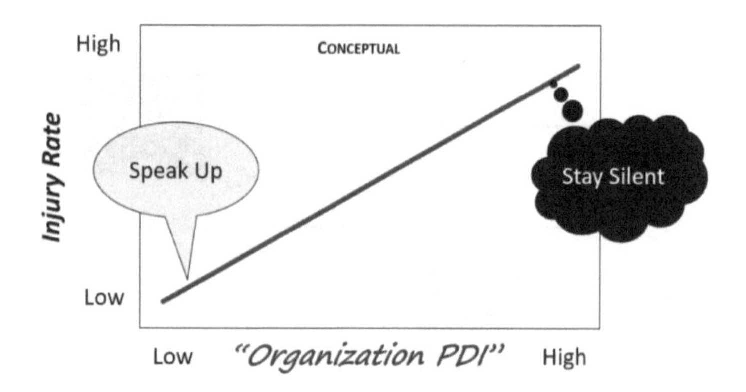

Based on the results of the survey question presented earlier, there is empirical evidence this relationship exists. There are many factors that contribute to a culture of commitment and a safe work environment. But if a significant proportion of employees stay silent, it is a symptom of an organization that manages safety mostly through compliance. In a work environment where blame and/or fear are common, there is little doubt silence will prevail.

Escalating concern

The airline industry made changes to address the identified cultural communication gap. Crew resource management training is now standard in the industry. It is designed to teach junior crew members how to communicate assertively.

There is a standard procedure for co-pilots to challenge the pilot if he/she thinks something has gone wrong or a poor decision is being made. It is a set of escalating statements:

"Captain, I'm concerned about…"
"Captain, I'm uncomfortable with…"
"Captain, I believe the situation is UNSAFE."

If the captain does not respond, the first officer is required to take over the aircraft.

South Korean people have a high Power Distance Index, and historically pilots from this country suffer a high plane crash rate. Since these changes in crew resource management training, South Korean commercial aviation crash rates are now in line with those countries with a low PDI (see the earlier graphic). Note South Korean culture still has a high PDI. But the pilots have been trained to behave quite differently once they enter the cockpit. They are now much more likely to speak up and question the captain if they see or hear something which could jeopardize the safety of the flight.

We could take a page from the crew resource training manual and apply it to an industrial setting. Why not give employees the skills to speak up using a standard protocol?

For example, the procedure for anyone to challenge a co-worker when he/she thinks that individual is taking an unnecessary risk could be the following:

"__________, I'm concerned about…"
"__________, I'm uncomfortable with…"
"__________, I believe the situation is UNSAFE."

If the co-worker does not respond, the task is stopped. Work does not proceed until someone else (usually a supervisor) is consulted.

Polite doubt by the lone voice

Just because we educate employees on how to escalate a concern, this does not mean they will have the *courage* to do so. We need to appreciate the powerful force of social influence. David Maxfield and Joseph Grenny have demonstrated the power of having just one person in a group speak up in dissent. They suggest one effective way to do this is to express disagreement using "polite doubt." If you disagree with a group, you can say something like, "Well, I guess I see things a little bit differently than you do. Here's what I think…"

Remember the discussion earlier on social pressure and the classic experiments by Asch on conformity (*Chapter 15*)? In another variation of his original experiment, Asch broke up the unanimity (total agreement) of the group by introducing a dissenting confederate or ally. He found even the presence of just one ally who goes against the majority choice can reduce conformity as much as 80%.

Recall in the original experiment 32% of participants conformed by giving the wrong answer. But when just one confederate or ally gave the correct answer using polite doubt, conformity to group pressure dropped to only 5%.

It starts with you

There are many case studies that support the notion when people in a group fail to speak up, the resulting outcome can range from unsatisfying (a project failure) to lethal (the Challenger disaster).

If there is a norm of not speaking up in your organization, it is critical as a leader you show a new norm for candor. This starts with your expectations. You create the beliefs people hold by the things you say and do. It doesn't take long for everyone in the group to get the message, "I am not interested in what you have to say if you disagree with me," if any contrarian view is dismissed or minimized. Only through repeated experiences of candor being positively recognized and encouraged can you hope to break an existing conspiracy of silence.

Don Merrell of JR Simplot wrote an especially poignant poem that addresses the notion people have a tendency to not speak up even when they know it is the right thing to do and it could save someone's life. His poem is reproduced on the next page.

I Chose To Look The Other Way

I could have saved a life that day,
But I chose to look the other way.
It wasn't that I didn't care,
I had the time and I was there.
But I didn't want to seem a fool,
Or argue over a safety rule.
I knew he'd done the job before,
If I spoke up, he might get sore.
The chances didn't seem that bad,
I'd done the same, he knew I had.
So I shook my head and walked on by,
He knew the risks as well as I.
He took the chance, I closed an eye,
And with that act, I let him die.
I could have saved a life that day,
But I chose to look the other way.
Now every time I see his wife,
I'll know I should have saved his life.
That guilt is something I must bear,
But it isn't something you need share.
If you see a risk that others take,
That puts their health or life at stake.
The question asked or thing you say,
Could help them live another day.
If you see a risk and walk away,
Then hope you never have to say,
I could have saved a life that day,
But I chose to look the other way.

Conclusion

We cannot achieve a zero event work place unless we create an environment where employees are watching out for one another. Having high situational awareness is simply not enough. We need everyone to be comfortable enough to take action and speak up when they see anyone taking an unnecessary risk - every time!

Refer to the **SAFETY LEADER'S TOOLBOX™** for suggestions on how we can get employees to speak up when they see risky behavior.

- **Set clear expectations** everyone (even the most junior employee) is empowered to speak up whenever something doesn't seem right.
- **Positively reinforce this behavior** when it is observed. Communicate the importance of doing this by citing examples or telling stories of others who spoke up and prevented a decision or action that otherwise could have resulted in injury.
- **Be a role model** receiving feedback. Publicly praise anyone who voices a concern. This is especially critical if someone expresses an opinion counter to the majority view.
- **Provide training** on how to raise and quickly escalate a concern, e.g., the crew resource training method used by the airline industry. Standardize the way employees can respectfully disagree or pose a question through polite doubt when there is a hierarchy that may inhibit the behavior to speak up.
- **Enlist Opinion Leaders** in the effort to make "speaking up" an accepted and expected behavior. Recall these are people who are "respected and connected" in the organization. If you can get this group to speak up, it sends a signal to others it is an acceptable norm.
- **Facilitate a discussion** about this topic in natural work groups. Have them commit to one another: (a) they will speak up (b) they will listen when anyone questions a decision or believes a situation is not safe (c) they will watch out for one another.

PART FIVE

The Power of Actions

"If you have to ask what it symbolizes, it didn't."

— Roger Ebert

17

SHOW AND TELL

Symbolism can be a powerful method to align an organization to a new way of thinking. Transformational leaders know everyone will be looking at what they say and do. One of the basic tenets of effecting culture change is to create experiences that will reinforce the beliefs you want people to hold. These experiences can be small words of support, a well-written sincere note, or a key policy decision. But they can also be bold acts that create a lasting impression and become the basis for stories and legends.

Consider the following symbolic acts and stories.

At the 1995 World Cup rugby final, Nelson Mandela put on a jersey of the South African Springboks, which under apartheid had been the exclusively white national rugby team. Mandela purposely chose to wear the uniform of the sport that black South Africans had always seen as that of the oppressor. The symbolism was unmistakable. It was an overt statement to millions of South Africans he sincerely believed in reconciliation in the new democratic South Africa.

Gordon Bethune, CEO of Continental Airlines, was trying to send a message to all his employees the old rigid rules were history. He wanted everyone to make whatever decision they thought was right for the company and the customer. To make this clear, he staged a book burning, where the previous "rules manual" was ceremoniously set afire in a 55-gallon drum in the parking lot. This story soon spread among the employees. Message received.

Meg Whitman tells the story of how she needed to change the culture of HP when she took over as CEO. The senior leaders had lost touch with their employees who had low levels of trust with their executive team. One of the first things she did was get rid of the executive parking lot at the headquarters which was actually surrounded by barbed wire and protected by security guards. In addition, all the executive offices were moved from closed-door wood paneled rooms to cubicles placed in the middle of the other administrative areas.

A legendary story from FedEx is held up as an example of what it means to guarantee any package could be delivered just about anywhere overnight. One of the company's drivers was out late one snowy night to check a drop-off box for any packages. Unfortunately when he got to the box, the lock was frozen solid, and his key broke off in the lock. After trying in vain to reach the packages inside, the driver finally made the decision to drive to a nearby auto garage. He borrowed a torch which he used to cut the legs off the box. The driver then put the box into his truck and delivered it to the airport. A maintenance team was able to drill the box open, remove the packages inside and get them on the plane. This story reinforces the message FedEx will positively do everything it can to get a package to its destination on time.

Perhaps my favorite example of a leader effectively using symbolism happened over a century ago. In 1914 Ernest Shackleton led a group whose goal was to cross the unexplored Antarctic continent.[68] In January of 1915 their ship, *Endurance*, was trapped by pack ice on the Weddell Sea. Nine months

---

[68] The incredible story of *Endurance*, led by Sir Ernest Shackleton, has been recounted many times. Not a single man perished, despite the crew of 28 traveling thousands of miles on ice and open water in dreadful conditions over many months. Their survival is largely attributed to the leadership skills of Shackleton.

later, Endurance sank under the crushing pressure from the ice.

Shackleton decided their only hope for escape was to head toward open water. His crew packed the life boats with their gear and loaded these on handmade sledges. The idea was to harness the men who would pull these sledges across hundreds of miles of pack ice. Shackleton was convinced this sledge march would only be successful if every nonessential item was discarded. He had to let his crew know anything that would not enable this expedition to be successful had to be left behind.

ROYAL GEOGRAPHICAL SOCIETY. INSTITUTE OF BRITISH PHOTOGRAPHERS

Shackleton issued an order that each man could only take along two pounds of personal gear. He then used powerful symbolism to demonstrate what he meant by traveling light and what was truly nonessential. Standing where everyone could see him, Shackleton reached into his parka and pulled out a handful of gold coins and a gold cigarette case. With a dramatic gesture, he tossed all of these items into the snow at his feet.

One of the expedition members later recalled,

> *"Naturally, after witnessing this action, which brought home to me at any rate the shifting values in life and the knowledge that there are times when gold can be a liability instead of an asset, we all discarded everything save the barest necessaries."*

The use of symbolism or stories is not only within the purview of political leaders, company executives, or famous explorers. Creating your own symbolic experiences is only limited to your imagination and the thoughtful consideration of the message you want to send.

Here is a personal story where symbolism was used to send a visible and powerful message.

I was working at a manufacturing site for several years when a new leader was brought in to manage the facility. While he was new to the site, almost all the managers and supervisors had been in their roles for at least 5 years. Many had been there for 15 years or longer.

Within the first week, the site manager (let's call him Sam) held a meeting with the management team and commented housekeeping in general was deplorable. He expressed his displeasure and instructed each manager to embark on an effort to clean up his area. A week passed by. Sam commented again he had not seen any significant improvement and asked for the clean-up efforts to be stepped up. Another week passed by.

One Friday morning, as the managers were gathered for a daily production meeting, Sam entered the room and announced, "Folks, instead of our usual meeting, I want you to put on your safety gear and follow me. We are going to take a tour." We looked at one another as we donned hard hats, ear plugs, and safety glasses. Then we left as a group, with Sam in the lead.

Sam led our group through the entire facility. We didn't just follow the usual pedestrian walkways. We visited basements, offices, shops, labs, control rooms, docks, road ways, and outside storage areas. We even traveled to the roof! Every 30 seconds or so Sam would stop and simply point at something. Each time he directed our attention to items upon which he had *spray painted his initials in red.* Sometimes it was a pile of debris. Sometimes it was a spare part. Other times it was shelving filled with opened, unmarked boxes. There were fans, pumps, bearings, motors, pulleys, rakes, buckets, blades, gloves, wire, bottles, drums, grease fittings, chairs, lamps, cables. You get the idea.

The bright red graffiti brought all this disorganization out of the shadows and into the foreground. It felt like we were seeing the letter 'A' in Hawthorne's *The Scarlet Letter.* The shame and guilt of the group were palpable. How had we gotten to the point where we had accepted this work environment?

When we completed the tour and came back to the meeting room, Sam asked the group, "Any questions?" There were none.

It was a painful experience for all of us.

Of course we received the message loud and clear. Over the ensuing weeks, the facility received a much-needed facelift. In hindsight it is easy to see now what we could not see then. But it took a radical approach in bringing this to light. We had become myopic and complacent. Over time we had accepted the unacceptable.

It should be no surprise this organization went on to achieve improved levels of performance in safety and quality. Poor housekeeping was only a symptom of the larger problems of low expectations and a lack of personal accountability. Because Sam set a higher level of expectations *(see Chapter 13),* everyone gained a renewed sense of purpose and direction which had been absent.

Conclusion

Do you want to make sure everyone in your car always wears a seat belt? Refuse to start the ignition until every passenger is buckled up.

Do you want to create an "open-door" policy? Take all the doors in the office area off their hinges and remove them.

Are you concerned employees are taking short-cuts and unnecessary risks to keep a production line running? Shut down the line and huddle the team together to discuss the importance of working safely.

The next time you want to communicate something to your group that is critical, you could write a memo, send an email, call a meeting, host a conference call, etc.

Or you could plan a memorable event using symbolism that will be impactful and long-lasting.

What message do you need to send?

- **Use symbolism to emphasize what's important.** Be thoughtful about the message you want to send. You should communicate using traditional media, then emphasize the main point of your message with a symbolic act.
- **Orchestrate simple ceremonies to cement important decisions, events, or policy changes.** For example, do you want to embolden your employees to watch out for one another? *(see Chapter 16).* Consider a personalized exchange of commitments among members of a natural work group where they pledge to watch out for one another's safety, and speak up if they see someone taking an unnecessary risk.
- **Tap into the local culture for symbolism.** Some of the most effective symbols are those associated with the local culture or are widely understood. One company leveraged the popularity of NASCAR racing to instill pride of workmanship in their workforce. The company became a minor sponsor of a race car. NASCAR photos and analogies were seen everywhere on bulletin boards. Employees were even encouraged to paint the forklifts in the color schemes of different race cars. Quality, productivity, and morale soared.

"You can observe a lot just by watching."

— Yogi Berra

18

FASTER AND SAFER

One of the foundational Lean tools is QCO (quick changeover) which is also sometimes referred to as SMED (single minute exchange of die). SMED includes a set of techniques which makes it possible to perform equipment set-up and changeover operations in less than 10 minutes. Not every changeover can be completed in this amount of time. However, any operation would benefit from using this Lean tool if there is a requirement for:

- a change in "lot" types
- a process or set-up change

I will use the term QCO as being interchangeable with SMED. Most of the time, the opportunity for implementing QCO in a process is driven by the need for greater flexibility, quicker delivery, better quality, or higher productivity. These are indeed significant benefits that are realized because this approach identifies and removes some of the eight sources of waste.

There is an equally significant benefit to assessing a process and implementing QCO: the resulting process changes often make setups simpler & easier; therefore faster *and* safer.

The following case studies demonstrate how using QCO principles can lead to work which is not only completed in less time, but is also safer.

Case One: The Frustrating Fabric

Every six to eight weeks at a large manufacturing facility, a major equipment changeover takes place. While the machine is down for maintenance, a crew of eight persons is assembled to remove a large, heavy, wet, worn piece of fabric. Removing the old fabric is a fairly straightforward task. Because the old fabric can be cut, it can be removed in sections without much difficulty. However, the new replacement fabric that needs to be installed in the same position has an 'endless' design that cannot be cut. There are many tight locations and constricted areas. The crew is forced to stand and pull in awkward positions to complete the task. The entire procedure can take as long as two hours (longer if things do not go well).

The process owner wanted to find a better way to get this job done. He sponsored a team which used QCO thinking to determine improvements. First, the team mapped the process. Then they identified and implemented numerous changes for a faster installation. These included:

- installing a new winch
- adding pulleys
- developing a new slide
- modifying the bolt types

As a result the new fabric can be installed in about *half the time.* With these changes the work was now faster *and* safer for the crew. The easier installation translated to a significantly lower risk of injuries like strains, cuts, or contusions. One crew member shared, "I wonder why we didn't make these changes earlier. This is still a tough job. But now it is less frustrating..."

Case Two: The Rapid Razor

Another QCO opportunity was identified with the objective of reducing the time to changeover between rolls of paper on a small processing unit. As part of this process, a person is required to splice together two webs of paper. The operator has to prepare the splice by making a specific cut across the paper web, then use a special two-sided adhesive tape to join the paper together. This procedure is done several times during a shift.

When the team first observed this task, it looked simple. With assistance from the operator, the team filmed the splicing operation. Later when they reviewed the video in slow motion, there was an audible 'gasp' in the room.

The team saw the operator performed a cutting motion so quickly no one had observed what was clearly visible on film. Twice he rapidly passed the razor blade within an inch of his hand! When the team showed him the video and asked him about this motion, his response was "I never realized I was bringing the knife that close to my other hand. I've done this job for years, and it's just a habit I developed. I have only nicked myself a few times."

Working with the crews, the team devised some simple changes to the procedure that not only saved a few minutes of time on each change, but also reduced the likelihood the operator would risk getting cut. In addition, the operators were provided with a much safer cutting tool than the retractable blade version previously used. The result was a faster *and* safer work method.

Case Three: The Bad Back Bolts

Sometimes a little support can make all the difference. When a team was charged with reducing the set-up time for an industrial chipper, their original goal was to gain an additional 15 minutes on each 100-minute changeover. The process was studied in detail. Some internal tasks were changed to external tasks. An additional 5S effort[69] reduced the time looking for tools.

However, the observation that caught every person's attention was the process used by the operator to remove (and later torque) over fifty bolts. In order to reach some of the bolts with a heavy air wrench, at times he was reaching over his head. Other times he was bent over at the waist. Even a novice could see his body positioning was a prime factor for potential back strains which some operators had reported over the years.

The solution? Someone retrieved a portable adjustable-height stool from an office area. The operator remained in a seated position with the air wrench at chest level. He was now in a more ergonomic position for this repetitive task. The chipper was then rotated to bring the bolts into a position where they could easily be reached. The new procedure was faster *and* safer. The operator commented afterwards, "This job doesn't seem nearly as tiring ..."

Pull, Tool, Stool

Three simple changes:

- An easier way to pull a heavy object
- A better tool to use for the job
- A stool for good ergonomics

Three improvements made a difference in reducing the risk of injury. In each case small changes made it easier, faster, *and* safer to do the job. And what was the investment? The intellect of a small team and the people who actually do the work.[70]

---

[69] The 5S methodology was previously defined on p. 105

[70] The Rapid Action™ process by Leap Technologies was the primary method used by the teams to engage employees in these continuous improvement opportunities. *https://www.improvefaster.com/*

Rushing vs Efficiency

I want to make a clarification on this topic. Some people confuse working faster with working more efficiently. It is important to note there is a distinction between rushing to get a job done and making the job easier to accomplish.

If someone feels rushed, they are more likely to take shortcuts and place themselves at risk *(see Chapter 3)*. In these cases usually no one has performed an assessment of how the work is done. The person is simply expected (often by the supervisor) to just do the current job faster. Although the operator may be able to complete the job more quickly, it is only because he is forced to pick up his pace in an inefficient process. But this speed comes with a price - risk. Indeed, the "rush factor" is a state of mind which has been proven to contribute to errors and increase risk.

If someone is performing a task where the job has been re-designed to make it easier, the work is more planned and methodical. As a result the job can be completed more quickly, as well as with less risk. The person is attentive to the task at hand. He is less likely to be fatigued or frustrated with how the work is getting done. He is faster because he is more efficient; there is less waste in the process.

Conclusion

QCO is a Lean tool used to focus on reducing the time it takes for equipment set-ups or changeovers. It has huge potential benefits in quicker delivery, higher quality, and increased productivity among others. These principles also result in improved work processes that are simpler and easier; therefore faster and safer. This aligns with the notion people will perform a task in a certain way if they have the ability to complete the task **and** if we make it easier to accomplish.

If we simply demand people work faster without providing an opportunity to lean the process, we introduce the "rush factor" into their work. This significantly increases their risk profile. On the other hand, if we are diligent in re-designing the work using QCO principles and removing waste, we gain productivity while decreasing risk.

- **Assess your processes for QCO opportunities.** If you have areas where you are changing over equipment to produce different products, you have a QCO opportunity. Process observations can reveal if there are sources of waste. Enlist someone who has experience with SMED or QCO to reduce waste and make the task easier.
- **Be aware that working quickly may or may not lead to taking more risks.** If the employee is working at a good pace because of an efficient work design, it is possible he is *less* exposed to risk of injury. However if the worker feels rushed because he is trying to meet a certain productivity quota and the process has a lot of waste, he is *more* likely to take risks. Are there any jobs your employees do that fit the latter description?
- **Solicit employees for ideas on how to make their work easier.** Not every idea suggested will be actionable. You will be surprised at the number of small changes that can be implemented. Many of these will not only improve productivity, but could also reduce the risk of injuries.
- **Use symbolism if necessary to make a point.** *(see Chapter 17)*. Most employees just want to do a good job. They want to do the right thing. Sometimes they take risks simply because they want to keep the process running. Step in and stop the process if you see someone making a choice to take an unnecessary risk driven by their perception (of productivity over safety). It will send a powerful message to everyone.

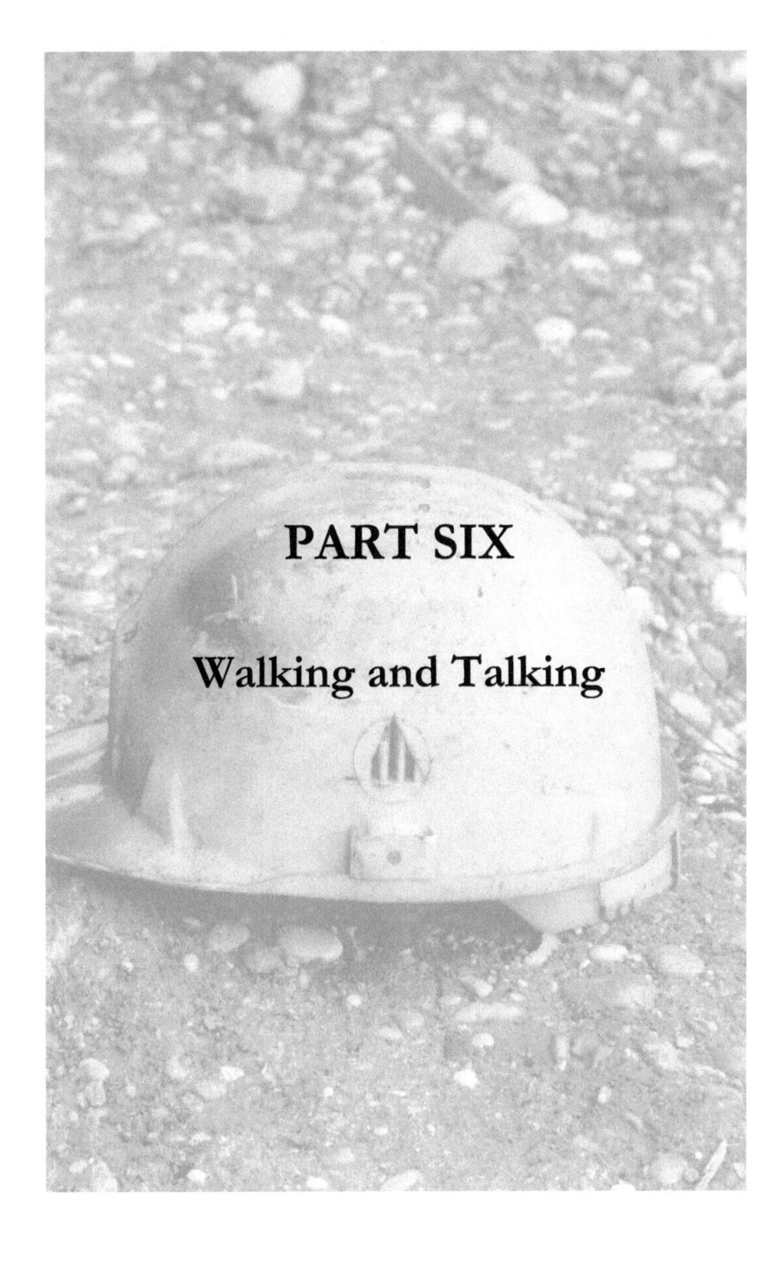

PART SIX

Walking and Talking

'There's a lot of difference between listening and hearing"

— G. K. Chesterton

19

WHY SHOULD I LISTEN?

Too often we view safety conversations as simply something that has to be done. We know these conversations are a responsibility of any leader. However, when we talk with employees about safety, our conversations are often reactive and seldom well-planned.

"The toolbox topic for today is chemical safety," the supervisor said as he looked down at the information sheet he was holding. "The first group of chemicals we will review are known as acids. Strong acids have a pH significantly less than 7. Some examples include..."

With this introduction the warehouse crew in the room immediately tuned out. Some looked at the ceiling. Some stared at their shoes. Some even took a quick peek at their cell phones.

If a safety meeting is conducted where the primary goal of the leader is to "check the box" for a required training, the response is predictable. When employees realize the objective, there is no engagement. Most attendees stare blankly or watch the clock. Group meetings such as these should not even be considered 'conversations.' They have almost no impact in terms of engagement, learning, or mindset.

There are three essential attributes for proactive safety conversations to have a positive impact. Think of these as the **3 P's** of an effective proactive conversation:

- **P**rincipled
- **P**revalent
- **P**ersonalized

Principled

In Chapter One I discussed the importance of starting every safety conversation with the right "why."

If the reason you have any safety conversation is because you care, your approach will be to coach and seek commitment through collaboration.

Your employees need to know you care about them. If their perception is you are going through the motions or simply satisfying a requirement, they will disengage.

You may indeed have to occasionally review rules or policies. Almost every organization has certain mandated safety topics. It makes a significant difference if you deliver these messages with a "why" of caring.

It does not matter if you hold a one-on-one safety conversation or if you talk to an entire work group. It does not matter if the topic of the conversation is mandatory or not. Regardless you are more effective, and your message has more impact when your motive is based upon a core principle of personal caring.

Prevalent

There are three situations when safety conversation topics can be considered prevalent: significant hazards, common risks, and potential error traps.

In many work environments hazards are everywhere. However, some hazards are more significant in terms of their potential to cause a serious injury. Identify tasks where there is a significant hazard. Proactive conversations with this focus could be the difference in whether someone is exposed to a hazard that results in a life-changing event.

Similarly, it is not unusual for people who perform certain tasks to take the same risk. Their decision to take an unnecessary risk could be rooted in a false perception they are not likely to get hurt. Or perhaps they are influenced to take this risk because something makes it difficult to do the job the safe way. Regardless, a safety conversation that probes the reasons why someone decides to take a risk can be revealing.

People are fallible. We make mistakes. Frequently we are exposed to error traps in the workplace (situations or conditions that make it more likely for us to make a mistake). Understanding error traps and collaborating with the employee to mitigate these are extremely effective in reducing the likelihood of human errors that can lead to injuries or incidents.

Personalized

Let's go back to the warehouse crew. Obviously, chemical safety was not meaningful to this group of employees. (Unless, of course, they were handling, transporting, or loading chemicals). The topic was not personalized.

The first question these employees asked themselves when the meeting started was WIIFM (What's in it for me?). As soon as they determined the answer to this question ("not much!"), they mentally checked out.

It is your responsibility to make a personal connection between the topic and your audience. If the safety conversation is not related to (1) a job they do or (2) a place where they are active, it will be difficult or impossible to keep their attention.

Principled, Prevalent, and Personalized

So which proactive safety conversations have the most impact? A principled and effective proactive safety conversation

always starts with a "why" of caring. It has the potential for high impact when (1) the topic addresses a significant hazard, risk, or error trap and (2) these situations are present in the work being done or the areas where the employee is engaged in an activity.

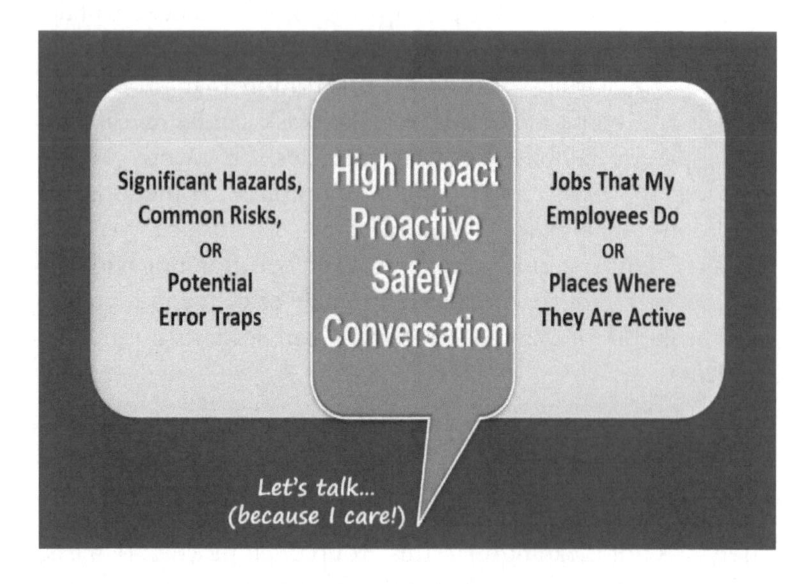

As an example, which one of these conversations is more likely to have a greater impact on a worker's behavior?

- Reminding the employee to wear their required PPE (personal protective equipment).
- Encouraging the employee to think about the ways (and the reasons) he could end up in the line-of-fire while performing a specific task.

You should not *avoid* having the first kind of conversation. However, these kinds of topics tend to be one-sided, reactive, and shallow. In addition, the message is frequently delivered from a compliance motive. (Even a reminder about PPE could be more impactful if you express genuine concern and make a connection to the employee's specific task or activity).

Compare the PPE example with the second example of a conversation. This one is open-ended. It invites a dialogue. It has the potential to engage the employee in a "What if…?" thought experiment. As long as it is facilitated from a perspective of caring, it is a **principled** conversation. This type of conversation focuses on hazards, risks, or error traps which are **prevalent** and is **personalized** to the employee.

If you aren't sure where to begin, consider existing safety rules, policies or procedural documents. Many organizations have completed JHA's (Job Hazard Analysis) or JSA's (Job Safety Analysis). These are rich sources for a potential discussion. Use these documents as starting points to engage your employees.

The figure below shows another framework used to select a topic for a proactive safety conversation. It combines a task which has inherently greater risk with multiple potential error traps (i.e., distraction, rushing, complacency, or fatigue).

List three "high-risk" tasks* where you believe an employee was almost certainly ***distracted*** at some time while he/she was performing this job.

	TASK #1	TASK #2	TASK #3
Place a "check mark" in the box for the **Error Traps** below that are *frequently present* for the person who is doing this task			
Rushing			
Complacency			
Fatigue			

**defined as a task where the consequences of making a mistake could result in a serious injury*

Because we are frequently distracted *(see Chapter 3)*, it is helpful to think about some "high risk" tasks where an employee is susceptible to any number of error traps. These situations are natural candidates for an impactful conversation, as long as they are delivered from a motive of caring and are relevant to the person who is doing the task.

Conclusion

Proactive safety conversations that are principled (based on caring), prevalent (address common hazards, risks, or error traps) and personalized (relevant to the individual) have the most impact.

Use these criteria to initiate and frame your conversation. You are more likely to have a discussion where the employee is engaged, and the conversation is memorable.

- **Ask your employees what risks or hazards they want to discuss.** Who knows best what situations or conditions are most likely to put them at risk? The people who actually do the work every day. Ideally, ask some of the crew members or associates to lead a discussion on a safety topic that interests them. Offer to assist them in preparing for the talk.
- **Use the "high risk task vs error trap" framework to identify safety conversation topics.** When I guide clients through an exercise using this tool, they are amazed at the number of relevant topics that surface. Importantly, the topics are specific to each process. The kinds of things that are most impactful in one work area are often very different from the topics that resonate with employees in another area.
- **Assess near miss and incident reports for opportunities and trends.** If someone was hurt (or nearly injured) in the facility, this topic should be discussed with all employees. Just be sure to make the connection between how the documented incident in one area is translated to a prevalent risk in another area. For example, was someone in the warehouse nearly hit by a forklift? Broaden your thinking to consider the kinds of *vehicle-pedestrian interactions* (not only those involving forklifts) that could occur in each work area.

"Humility is not thinking less of yourself, it's thinking of yourself less"

— C.S. Lewis

20

ARE YOU ASKING OR TELLING?

Reflect for a moment on your communication style as a leader. Do you spend more time **asking** or **telling**?

In the United States we have a culture of "do" and "tell". We value task accomplishment more than relationship building. It is a cultural bias many of us have. A majority of people in a supervisory or managerial role spend a significant amount of time telling others what they *think* their employees need to know to get the job done rather than asking for their input.

Status in most workplaces is gained by task accomplishment. We are recognized and rewarded for getting things done. Indeed, one of the most significant factors that determines whether someone is promoted or given more responsibility is the ability to complete work assignments.

In some cases a mostly "tell" approach is all that is required to get a task accomplished. Some examples include situations where the work is straightforward or where the employee is inexperienced. These interactions are characterized almost exclusively by one-way communication. While we can get work done using a telling approach, it does very little to build relationships.

There is a significant amount of interdependence in today's workplace. We need effective communications and good relationships to be successful in completing complex tasks. To build these relationships, we need a different technique other than simply "telling."

It is helpful to understand the differences between a strategy of telling and one centered on a specific kind of asking known as *humble inquiry*. We will learn this attitude is critical for improving communications, developing relationships, and building trust.

The notion of humble inquiry was originally developed and explained by Edgar Schein.[71] He defines humble inquiry as:

> *"...asking questions to which you do not already know the answer; building a relationship based on curiosity and interest in the other person."*

The strategy of asking by using humble inquiry is especially critical for facilitating an effective safety conversation. Many safety conversations initiated by people in authority are missing this key attribute. Unless you are willing to ask questions from a basis of genuine care for the employee, you will not build trust.

More TELLING than Asking

Figure 1 depicts a model of a common management strategy where we do more telling than asking. It is primarily **task-oriented.**

What is your principal motive for having a conversation? For supervisors who align with this management strategy, their motivation is often driven by the need to maintain or exert control. Often the primary goal of these conversations is to accomplish a task with a secondary goal of getting the job done while being compliant with procedures, rules, standards, or policies.

---

[71] *Humble Inquiry*. Edgar H. Schein. Berrett-Koehler Publishers. San Francisco. 2013.

Figure 1

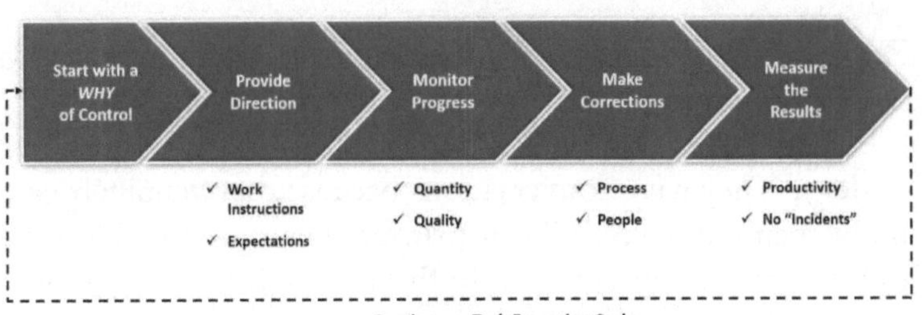

Continuous *Task Execution* Cycle

A supervisor often defaults to a "telling" strategy simply because he or she has the authority or doesn't have the time. Directions are given via work instructions. Expectations are set in terms of specific job duties and outcomes. Progress is monitored by the amount of work being accomplished and the quality of the output. Feedback is given in the form of corrections to any work behaviors. Modifications are made to the process as needed. Finally productivity is measured, and checks are performed to see if compliance with safety or quality standards was achieved.

The entire cycle is repeated for the next series of tasks. Success is measured by work output and the absence of any deviations from standard. Productivity was acceptable. Quality was achieved. No one was injured.

Anyone with authority can successfully use this strategy if their primary goal is to accomplish tasks. However, it has severe limitations if you need to build relationships or trust. The reason: conversations that occur throughout this cycle are mostly directive.

This approach is especially deficient when it comes to safety conversations, where the primary objectives are to encourage open discussion and increase engagement. Let's take a look at a strategy designed for relationship and trust building.

More ASKING than Telling

Figure 2 depicts a model of a leadership strategy where we do more asking than telling. It is primarily person-oriented.

Leaders who start with a "why" of caring tend to use this strategy. They have conversations because they genuinely care about their employees. Their **primary** goals are to build trust and to increase engagement. In this case task accomplishment is a secondary goal. They understand if the right kinds of relationships are developed, the work will be successfully and safely completed by engaged and committed employees.

Figure 2

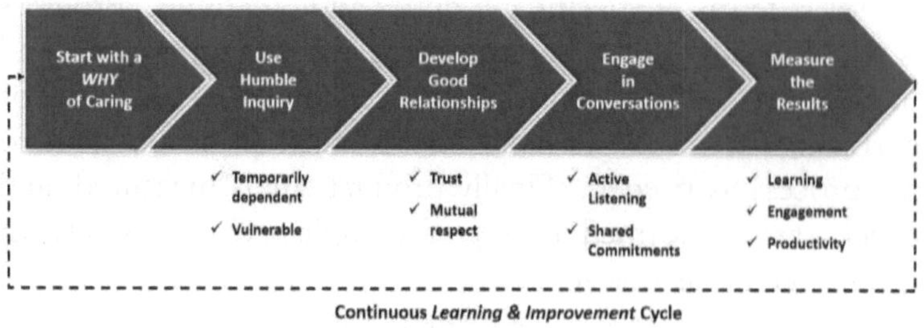

Start with a *WHY* of Caring
Use Humble Inquiry
Develop Good Relationships
Engage in Conversations
Measure the Results
✓ Temporarily dependent
✓ Vulnerable
✓ Trust
✓ Mutual respect
✓ Active Listening
✓ Shared Commitments
✓ Learning
✓ Engagement
✓ Productivity
Continuous *Learning & Improvement* Cycle

The leader who uses this strategy leverages the concept of humble inquiry to frame the conversation.

Several examples of humble inquiry are provided in the first step of the Five-Step Guide to a Safety Conversation[72] which will be discussed in the next chapter. They include sincere questions such as:

> *"Can you help me understand why...?"*
> *"Can you think of the major risks with this task?"*
> *"What mistakes could be made when doing this job?"*

Schein describes humble inquiry this way:

---

[72] p. 222

> *...it goes beyond just overt questioning. The kind of inquiry I am talking about derives from an attitude of interest, caring, and curiosity. It implies that [we are willing to make ourselves] vulnerable.*
>
> *The kind of humility that is crucial for the understanding of humble inquiry is* **"here-and-now humility."** *This is how I feel when I am dependent on you. My status is inferior to yours at this moment because you know something or can do something that I need in order to accomplish some task.*

If I feel I have something to learn from you or want to hear some of your experiences because I care for you or need you to accomplish a task, this makes me temporarily dependent and vulnerable. It is precisely my subordination that creates 'psychological safety' for you and therefore increases the chances you will tell me what I need to know. If you exploit the situation or take advantage of me by giving me bad information, I will learn to avoid you in the future or punish you if I am your boss. If you tell me what I need to know and help me, we have begun to build a positive relationship.

The leader's motive of caring and his sincere humble inquiry sets the stage for a relationship that can be built upon trust and mutual respect. One conversation won't get you there, of course. However, it starts a healthy process which includes active listening and shared commitments between the supervisor and the employee.

The key measures of success for this strategy are how much learning is taking place and the degree employees are engaged. One indicator of engagement is the extent of collaboration on any improvement ideas that emerge.

As illustrated in *Figure 2*, **each personal conversation motivated by caring and framed with humble inquiry contributes to a virtuous cycle of learning and improvement.** Productivity is not the primary goal of this communication strategy. Ironically, the quantity and quality of work almost always improve with the resulting positive relationships, trust, learning, and engagement.

Humble inquiry is used frequently by those who practice servant leadership. Asking your employees these kinds of questions sends a strong signal you care about them. There is a strong inference, "If you help me, then I can help you." And helping each other is a key building block of a trusting relationship.

Conclusion

Most of us spend more time telling than asking. While it may be appropriate to mostly "tell" in certain situations, telling does nothing to foster positive relationships or build trust.

When engaged in a safety conversation, it is critical to do more asking than telling.

By using humble inquiry as our approach, we become temporarily dependent because we want to learn something or the other person can do something we need. We also become vulnerable. We can be taken advantage of, ignored, or given bad advice. But this vulnerability or self-exposure also makes the relationship more personal. It shows the other person you are willing to invest a part of you to develop the relationship. This is how we start to build trust. And trust is essential if we seek collaboration and commitment.

The next time you engage in a personal conversation consider what you want to achieve. If you want to go beyond getting a task accomplished and develop a relationship built on trust, choose the right approach.

Asking through humble inquiry is superior to **telling.**

- **Reflect on your prevalent communication strategy for direct reports.** Are you mostly in a directing mode? Is your primary goal task accomplishment? Make a conscious shift to doing more asking than telling, especially when you are engaged in safety conversations. When personal interactions are initiated from a motive of caring, this approach is the starting point for building trust.
- **Practice here-and-now humility.** Unless you have far more experience than the person who is doing the work, you need their help to be successful in accomplishing many tasks. Have the courage and humility to become temporarily dependent on the other person and ask for help.
- **Hone your listening skills.** After you ask, listen! Most of us listen with the intent to reply. Here are some tips on how to be a better listener:
    - Remove all distractions. Be "in the moment" while having a conversation.
    - Listen to the speaker's signs and sounds. Be aware of cues from their posture and body language.
    - Repeat back what you hear in your own words. This tells the other person you are an *active listener*.
    - Leverage silence to encourage more dialogue. Resist the urge to immediately fill any silent voids by talking. Let your conversation partner finish his or her thoughts before speaking.

"One good conversation can shift the direction of change forever."

— Linda Lambert

21

THE **RIGHT** CONVERSATION

Most of us believe we are good at communicating with other people. After all we do it all the time, every day. Unfortunately, it is my experience when it comes to having personal safety conversations, we simply aren't very skilled.

For a conversation to be effective, each person needs to alternately talk and listen. Unfortunately, some leaders are prone to lecturing with very little listening. This ineffective communication style isn't isolated to senior leaders who ascribe to the command-and-control approach to management. It can be seen at all levels of organizations.

The prevalent communication style of managers and supervisors is a barometer of the safety culture. Reactive and occasional one-way safety conversations are telltale signs of a culture of compliance. Frequent interactive safety conversations are indicative of a culture of commitment.

A fundamental premise I proposed earlier states your personal motive for having any safety conversation significantly influences your safety culture[73].

- If the reason you have any safety conversation is to exert control, the approach will be to criticize and seek compliance through correction.

[73] pp. 2-6

- If the reason you have any safety conversation is because you care, the approach will be to coach and seek commitment through collaboration.

One communication model suggests an effective organizational conversation has four attributes: intimacy (building trust and listening), interactivity (promoting discussion), inclusion (collaborating on solutions), and intentionality (sharing a common purpose).

In this chapter I introduce a *guide* for an effective safety conversation - one that starts with caring. This guide incorporates attributes of an effective conversation. It also includes some of the key safety leadership concepts discussed earlier. This approach stimulates a conversation which enables coaching and collaboration.

Five-Step Guide for a Safety Conversation™

FRAME THE CONVERSATION ➡ ASSESS RISK-TAKING ➡ DISCOVER ERROR TRAPS ➡ IDENTIFY THE BEHAVIOR ➡ TAKE ACTION

It is noteworthy this framework can be applied whether the conversation is *reactive* (after an incident) or *proactive* (identifying factors that may contribute to errors or injuries). Let's walk through each of the five steps and discuss them in more detail.

FRAME THE CONVERSATION

When you initiate a safety conversation with an employee, the first few moments set the stage for what follows. Your intent will be clear to the listener by your body language, your words, and your tone.

If your words indicate you need control, then you are not having a conversation. Instead you are signaling your intent to deliver a one-way message.

"We've got a problem"
"You know the safety rules"
"I expect you to follow them"

In this case the employee immediately feels defensive. He is likely to share only the minimum amount of information.

On the other hand, if you frame the conversation with care and concern, the discussion is much more likely to be interactive. You do this by asking questions in a non-threatening way. These questions (and the way they are asked) are rooted in *humble inquiry.*[74]

"Can you help me?"
"What is the major risk?"
"What mistakes could be made?"

By asking questions like these, you are making it clear you need their help. The employee is being encouraged to participate in a genuine dialogue.

ASSESS RISK-TAKING The only way you can learn about hidden process or organization issues is to ask the right questions in the right way and then *listen* for risk-taking factors. Most of us believe we are good listeners. But research shows we seldom hear what people are saying in the way it was intended to be heard. Active listening is hard work! *(See the* **SAFETY LEADER'S TOOLBOX™** *at the end of Chapter 20 for listening tips).*

Of course if your original motive is control, you are not listening - you are telling. You are stating your case either for what was done incorrectly or what rule or policy was not followed.

By contrast, an effective safety conversation includes listening for the responses to all questions. You are seeking to learn about any potential risk-taking factors. People don't take

---

[74] p. 214

risks without a reason. As someone once said, "People do what they do because it made sense to them at the time they did it."

Your challenge is to find out why it made sense to the employee to take an unnecessary risk.

If the conversation takes place after an incident or near miss, listen for indications one of the prevalent risk-taking factors was present when an employee took an unnecessary risk. If you are having a proactive conversation, inquire what might persuade someone to take a risk when performing a specific task.

DISCOVER
ERROR TRAPS

For a supervisor who wants to maintain or exert control, rules and policies are edicts. If an incident occurs, the first thing he wants to know is the procedure, rule, or policy that was not followed. If someone gets hurt, it almost always means one of these was violated. In a compliance world, many people believe if you just follow the rules, you won't be seriously injured. As proven earlier, this belief is simply not true.

If caring drives the conversation, a supervisor knows another critical part of listening is to discover potential error traps. These are conditions or circumstances that make it more likely for someone to make a mistake. A few examples are listed below:

- ✓ Time Pressure
- ✓ Distraction
- ✓ Vague Guidance
- ✓ Multiple Tasks
- ✓ Complacency
- ✓ Peer Pressure

These error traps may emerge as part of the conversation. Or a supervisor may find additional ones when he takes a holistic view of the situation. Either way, acknowledging these error-prone conditions is the first step in finding ways to mitigate their effects.

IDENTIFY THE BEHAVIOR Let's do a quick review of the key aspects of a Just Culture[75]. One definition is "a culture where failure/error is addressed in a manner that promotes learning and improvement while satisfying the need for accountability." In a Just Culture, three possible behaviors may contribute to an undesirable outcome:

- Human Error
- At-Risk Behavior
- Recklessness

A supervisor whose motive is control often assumes the primary reason for a safety incident is (a) someone made a mistake or (b) someone made a poor choice. With this mindset, it is not a surprise when the supervisor attempts to "correct" the behavior through some kind of admonishment.

If human error is involved, the employee may hear something like, "Pay more attention" or "Keep your mind on the job."

If an employee is deemed to have made a poor choice, he may be challenged to "Think before you act next time" or even "Just don't do stupid things!"

In contrast, a supervisor who believes in a Just Culture realizes more than 90% of the time people are 'set up' to make a mistake[76] (by error traps) or are likely to take a risk (if one or more risk-taking factors are present). This supervisor realizes truly reckless behavior[77] rarely happens.

His focus in a safety conversation is to actively listen for risk-taking factors and error traps. The employee's behavior is

---

[75] pp. 55-66

[76] *Out of the Crisis.* W. Edwards Deming. Massachusetts Institute of Technology. Cambridge, MA. 1982.

[77] pp. 62

identified as reckless only when there is a choice to consciously disregard a substantial and unjustifiable risk.

Reckless behaviors exist in the industrial world, but they happen only occasionally.

TAKE ACTION A supervisor who expects to maintain control takes actions which are directed toward compliance. These are typically some form of re-training, warning, counseling, or discipline. In a culture of compliance, supervisors assume people should perform work using standard procedures and abide by the safety rules. If they do not, then they need to be held accountable.

It is a mistake to simply warn, counsel, or discipline someone for not using a procedure or following a safety rule without understanding the reason for their decision. There could be hidden organization or process issues that contribute to risk-taking. For example, procedural drift[78] could contribute to employee non-compliance.

In a culture of commitment, a safety conversation concludes on a very different note. Because the conversation proceeds in a spirit of learning and co-discovery, the actions are built on collaboration.

When a mistake is identified, the supervisor seeks to understand and mitigate any error traps.[79] In some cases, this could include a collaboration on mistake proofing solutions. These are simple process design changes that make it easy to do the right thing and difficult to do the wrong thing.

If an employee drifts into an at-risk behavior, the risk-taking factor determines the best course of action. Coaching is often effective for inaccurate perceptions or risk-taking habits. Like addressing error traps, a collaborative effort is a proven approach to making it easier for someone to perform a task safely.

---

[78] pp. 44-46

[79] pp. 24

Counseling or Discipline

When is this kind of leadership action appropriate? Here is a general guideline:

- Someone has been given feedback multiple times
- Resources have been provided to perform the task safely.
- No error traps have been identified

...but their performance is still unsatisfactory

Or

The employee exhibits *reckless behavior* (the choice to consciously disregard a substantial and unjustifiable risk).

Carefronting

Theologian David Augsburger coined the term "carefronting" decades ago as a way to think about effective methods for conflict resolution.[80] He defines it as:

> *The skill of* **caring** *enough about oneself, others, and desired goals to* **confront** *inappropriate behavior responsibly, while offering the opportunity for change.*

I prefer *carefronting* over 'counseling' in a safety context. Very few supervisors are trained as professional counselors. Yet, if we accept the premise everyone can and should care about their fellow employees, then we all have the obligation to confront unsafe behaviors.

When faced with repeat or reckless behavior, rather than counseling the individual, consider an approach to *carefront*

---

[80] *Caring Enough to Confront* by David Augsburger. 3rd Edition. Revell. Scottdale, PA. 2009.

them. Why? Because you **care** so much about them and their fellow employees you are not going to permit any behavior that places them or others at risk of being injured.

Conclusion

Using this Five-Step Guide,[81] anyone can conduct an effective safety conversation. It is designed to be used when your primary objective is to promote learning and improvement.

This guide is centered on the "Why" of caring. When you start with this motive, the conversation that follows is more open. The real reward in having a candid dialogue is in discovering the hidden weaknesses in the process and in the organization. Making these visible is essential. It sets the stage for collaboration to improve the work processes and to eliminate sources of failures or errors.

Supervisors and managers need to be skilled in facilitating effective safety conversations. Having a daily proactive safety dialogue like the one outlined here is a cornerstone in building a culture of commitment. This conversation should be an integral part of your safety strategy *(see Chapter 2)*.

What will you learn and improve as a result of your next personal safety conversation?

[81] For more information on how to obtain a Pocket Guide for a Safety Conversation™ or how to download the free mobile app, visit my website: *https://www.continuousmile.com/safety-conversation-guide/*

- **Start every conversation with a Why of caring.** Make it clear to the employee you are talking to them because you care about them and their personal safety. This is true for proactive and reactive conversations. It is also true regardless of the situation – even if you think discipline is a likely outcome. (Remember carefronting)?
- **Resolve to have more <u>proactive</u> safety conversations.** Constructively challenge people who do the work to think about *what could go wrong* or where a mistake could lead to injury on tasks they do every day. Use humble inquiry to frame these conversations.
- **Don't prematurely judge the choices of others.** Until you have a conversation where your goal is to understand potential risk-taking factors and error traps that may have contributed to their decision, you don't really understand what they were thinking. It is essential you engage them in a way that encourages an open and honest dialogue. Only then are you able to take the appropriate action.
- **Close every conversation with commitments.** Unless you and the employee close with some kind of action, commitment or agreement, nothing is likely to change. Without this closure, the content of what was discovered or discussed is quickly lost or forgotten.
- **Document and track your conversations.*** If you don't measure these, you can't improve upon them! *(see the next page for a tracking tool).*

*The **PRO** version of the **Safety Conversation Guide** gives leaders the ability to quickly & easily document individual and group safety conversations. It features:

- Subscription-based service
- Custom data fields for each organization
- On-demand report generation
- Linked web and mobile applications
- Real-time information

https://www.continuousmile.com/safety-conversation-guide/

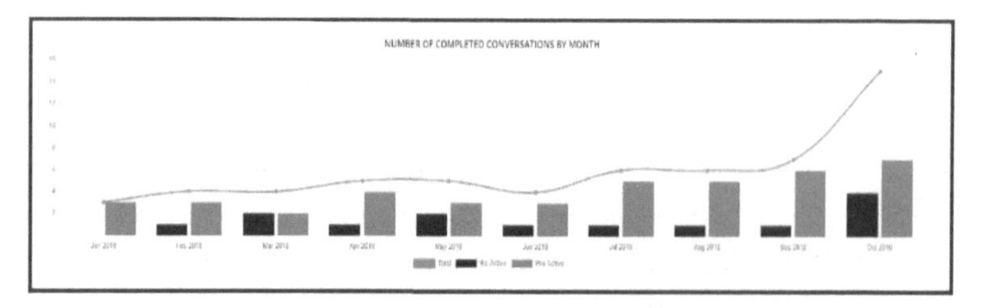

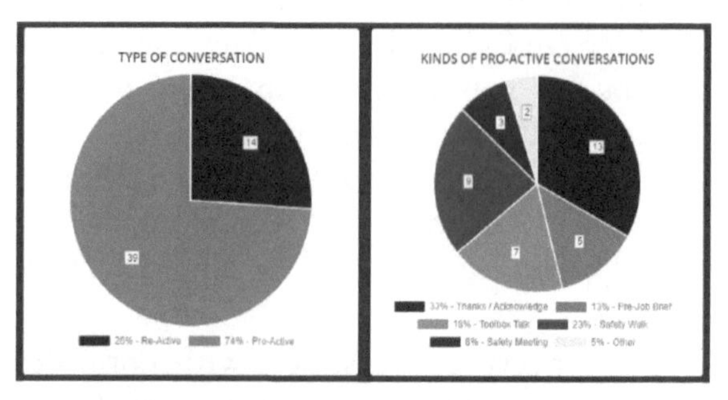

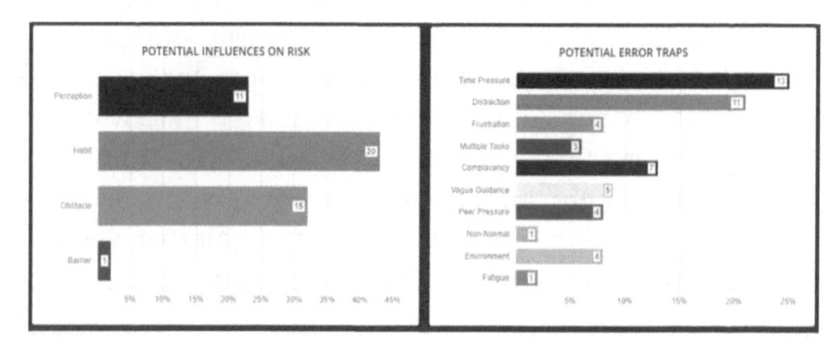

PART SEVEN

The Continuous MILE

"Never believe that a few caring people can't change the world. For indeed, that's all who ever have."

— Margaret Mead

22

DO YOU CARE?

Perhaps you are wondering, "How can I possibly remember all of these things well enough to be an effective safety leader?"

You don't have to be proficient in every aspect of leadership discussed in this book. Many of the recommendations in the **SAFETY LEADER'S TOOLBOX™** are designed to enhance your skills beyond basic competency. I firmly believe if you understand and embrace a few key concepts or principles, you are already well on your way to making a difference in the lives of others when it comes to safety. Think of these as building blocks for anyone who aspires to positively influence others.

Start with a "why" of caring. Begin every safety conversation by asking yourself why you are talking with the individual. *The primary reason should always be you care about them.* There are no exceptions. If you start your conversations with this motive, you will certainly conclude your discussion in a better place than if you focused on answering "how?" or "what?" It is the only way you can create a culture of commitment.

Facilitate frequent proactive conversations. We are not mind readers. The only way to know what someone *is* thinking (or *was* thinking) is to have a conversation. Using humble inquiry, ask them to share why they are (or were) doing a task a certain way. We may discover false perceptions or risk-taking habits we can proactively address. Or there may be something preventing them from performing the task in the safest way.

Build trust by fostering a Just Culture. The overall goal in this kind of environment is to learn and improve. All of us are fallible. We all make mistakes. We are easily influenced to take unnecessary risks. The actions of a leader who understands and appreciates these facts will reinforce a cycle of trust with his or her employees. This leader considers he may have a *situation* problem rather than assuming he has a *people* problem. The result is the best kind of accountability: one that discovers true root causes and identifies solutions to minimize future errors and risk-taking decisions while not allowing reckless behavior.

Be thoughtful with your words; be action-oriented. For anyone in a leadership role, words matter. And actions matter more. Choose words associated with the behaviors you expect. Use priming to reinforce the actions that will subconsciously influence others to take less risk and have more situational awareness. Back up your words with personal commitments. Get commitments from your employees.

Look for what is right and acknowledge it. If your observations and feedback are concentrated on what people are doing *wrong*, you are missing a huge opportunity to change your safety culture by reinforcing desirable work habits. Plan a safety walk and seek out examples of safe work behaviors or habits. There are many more of these than risky behaviors! Once you observe something positive, tell the person, "Thanks." By the way, your "Thank You checkbook" has unlimited funds.

A Conversation That Made a Difference

I have facilitated over one hundred workshops where participants learned about and practiced having the right conversations using the *Pocket Guide for a Safety Conversation*™.

While the workshop feedback is overwhelmingly positive, it is even more satisfying when someone from the workshop writes to let me know how they applied their learning. I received an email from Cody, a supervisor who attended my workshop. His note is reproduced verbatim here.

> *"We received a shipment of aluminum coils. (Each coil weighs about 5,000 pounds). They were shipped on a flatbed as opposed to the normal box truck. This meant they could not be unloaded in the traditional manner. We had to unload them off the side of the truck; we weren't sure if we had the right equipment to do this.*
>
> *The entire situation was very non-normal. We unloaded the coils safely by obtaining back-up and assessing the situation as a team.*
>
> *I had been fairly stirred by it because I felt without my input this crew would have rushed into the job, possibly making a critical error that could lead to an injury.*
>
> *I had a safety conversation with the crew who unloaded the coils. We talked about proactive forklift safety, abnormal circumstances, and how valuable it was that we put a team together before moving forward.*
>
> *They committed to calling for back-up the next time an abnormal situation presents itself, especially involving heavy equipment.*
>
> *I committed to looking into how we could make the whole receiving process for coils more formal, understood, and robust. I thanked them for calling me and for working together as a team."*

When I first read this email, my heart swelled with pride and my eyes filled with joyful tears. Cody "got it." He had taken what he had learned and facilitated the **right** conversation. Consider the various aspects of the situation and how each was effectively addressed:

- *we weren't sure* ***if we had the right equipment***... The resources required for the job were not available.
- *The entire situation was very* ***non-normal***... This is one of the error traps.
- *I had been fairly* ***stirred*** *by it*... Cody obviously cares about his crew.
- *this crew would have* ***rushed*** *into the job*... Another recognized error trap.
- *We talked about* ***proactive*** *forklift safety*... He facilitated a conversation about what could go wrong and what they could do about it.
- ***They committed*** *to*... Cody got a commitment from his crew related to a specific situation.
- ***I committed*** *to*... Cody took ownership for an action focused on learning and improvement (hallmarks of a Just Culture).
- ***I thanked them*** *for*... He acknowledged what the crew had done right and thanked them for their efforts.

There is no doubt Cody is making a difference in the lives of these crew members. Beyond this singular event, he set the stage for more fruitful conversations and discussions in the future. He planted a seed with each person on this crew to think about what might happen in other situations. He is building a trusting relationship.

Conversations like the one Cody had with his crew are essential for any organization which aspires to foster a Just Culture of Commitment.

Think back to the discussion of risk-taking factors and error traps. Recognizing these factors is an important part of nurturing an environment of learning and improvement. The challenge is many times these things are hidden from view. The people who know best what set them up to make a mistake or take a risk are those who actually do the work.

Using the iceberg as a metaphor, the things we can observe (incidents, interventions, and reported near misses) are above the waterline. But these outcomes are only a fraction of the thousands of precursors which lie below the waterline (risk-taking factors and error traps).

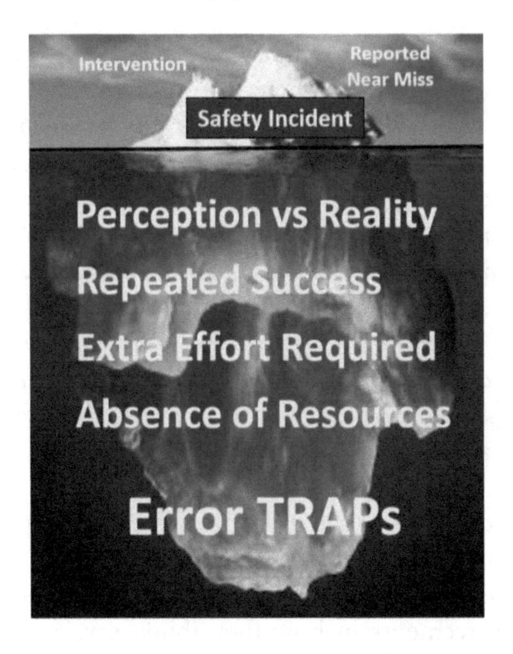

The only way to make these things **visible** is to have the right conversation. This is a conversation rooted in caring, framed with humble inquiry, focused on identifying risk-taking factors and error traps, and closed with actions and commitments. It is the kind of conversation Cody facilitated with his crew.

The Continuous MILE

When I started my consulting practice a number of years ago, one of the hundreds of decisions I had to make was the name. Because my industry experience included significant time as a Lean Six Sigma practitioner, I wanted to include the concept of **continuous improvement**. Safety is a complex set of processes that should be continuously improved.

The other part of the name for my business emerged as I considered how the most successful organizations achieved safety performance worthy of benchmarking.

Recall this excerpt from the *Preface:*

> *…over these years I observed quite a few leadership actions that significantly contributed to less risk-taking, greater hazard awareness and genuine collaborative efforts among employees and supervisors. Leaders who understood, embraced, and implemented these strategies saw a dramatic reduction in incidents and injuries at their facilities.*
>
> *In my experience, organizations that have the best safety performances do not have a secret unknown to poorer performing organizations. They simply do a lot of small things collectively and strategically well.*

The second half of the name for my consulting practice is an acronym: **MILE**. It stands for **M**inor **I**mprovements, **L**arge **E**ffect. This reflects my belief if leaders continuously make minor improvements in how they think, say, and do things, these changes can have a large effect on their organization's safety performance.

Indeed, in this book I have suggested **over 70 small ways you can significantly impact your safety culture.**

The Challenge

I hope you have been inspired to apply the safety leadership concepts outlined in this book. *Always* use the five basic principles outlined at the beginning of this chapter:

- Start with a "why" of **caring**
- Facilitate frequent proactive conversations
- Build trust by fostering a Just Culture
- Be thoughtful with your words; be action-oriented
- Look for what is right and acknowledge it

Build on your core competency as a safety leader by selecting tools from the **SAFETY LEADER'S TOOLBOX™**.

Decide on what minor improvement you can make in your leadership approach that will have the largest effect. If you have determined the largest gap in your safety culture by conducting a survey, use a tool that will help to close this gap. These tools are also valuable as you consider how to convert your safety strategy into an action plan.

The best way to engage in a safety conversation is to go where the work is actually done. Lean practitioners call this place the *gemba.* After all, nothing will change unless you take action. When you proactively go to the *gemba* with the specific mission to have a safety conversation, you are taking a **safety walk.**

A central theme of this book has been the critical role of our conversations in determining safety performance. Subtle changes in the way you think about risk and how you facilitate conversations can be difference–makers. You have gained significant insights into how to have an effective **safety talk.**

Starting today I challenge you to walk a MILE. Tomorrow walk another MILE. The following day walk another MILE.

This is your starting point for a rewarding safety journey. Like any journey whose destination is on the far horizon, it takes a consistent and determined effort. And it requires leadership.

Do you **care enough** to walk a **Continuous MILE?**

APPENDIX

Benchmarking performance of organizations that completed the Continuous MILE safety culture survey

Q	Leadership Dimension	Metric	Lowest Performance	Median Performance	Highest Performance
1	Personal Conversations	*Less than a week*	14%	55%	84%
2	Proactive vs Reactive	*Several times per week*	17%	48%	79%
3	Coaching vs Criticism	*Mostly a coach*	40%	61%	88%
4	Improvement Mindset	*Frequently encouraged*	30%	54%	92%
5	Active Risk Reduction	*Many ideas*	11%	44%	79%
6	Caring vs Controlling	*Frequently or always*	28%	74%	96%
7	Speaking Up	*Comfortable*	49%	76%	98%
8	Self Efficacy	*Confident*	66%	85%	97%

Survey data from 5500 employees representing 42 manufacturing organizations (2015 – 2018)

Chapter 1: Compliance, Commitment, & Leadership

1. *Start With Why. How great leaders inspire everyone to take action.* p39. Simon Sinek. Portfolio/Penguin. London. 2009.

2. *Control or Caring? What is your motive for a safety conversation?* David A. Galloway. EHS Today. March 14, 2016.

3. To learn more about the *safety culture survey* described here, send an email to: info@ContinuousMile.com.

Chapter 2: Developing a Safety Strategy

1. *2017 America's Safest Companies.* Sandy Smith and Stefanie Valentek. EHS Today. November 7, 2017.

2. *Influencer. The Power to Change Anything.* Patterson, Grenny, Maxfield, McMillan, & Switzler. McGraw-Hill. 2008.

3. *http://captology.stanford.edu/projects/behavior-wizard-2.html. B.J. Fogg. Stanford Persuasive Tech Lab.*

Chapter 3: Mistake Makers & Risk Takers

1. *Error Elimination Tools™*. Practicing Perfection Institute, Inc. www.ppiweb.com 2015.

2. *The Myth of Multitasking: How "Doing It All" Gets Nothing Done.* Dave Crenshaw. Jossey-Bass. 2008. https://davecrenshaw.com/downloads/multitasking-exercise-v2.pdf

3. *This is your brain on multitasking.* Garth Sundem. Psychology Today. February, 2012.

4. *Mistake-Proofing. Designing Errors Out.* Richard B. Chase and Douglas M. Stewart. CreateSpace Independent Publishing Platform. 2008.

5. *Mistake Proofing Reference Guide.* Quality Training Portal. 2009.

6. *Mistake-Proofing for Operators: The ZQC System.* Productivity Press. 2010.

7. John Grout's Mistake Proofing Center. *http://www.mistakeproofing.com/index.html*

8. *Understanding Influences on Risks: A Four-Part Model.* Terry Mathis and Shawn Galloway. EHS Today. February, 2010.

9. *Inside Out. Rethinking Traditional Safety Management Programs.* Larry Wilson & Gary Higbee. Electrolab Limited. 2012.

10. *The Field Guide to Understanding Human Error.* Sidney Dekker. Ashgate Publishing. 2006. ISBN 978-0-7546-4826-0.

11. *Blink. The power of thinking without thinking.* Malcolm Gladwell. Back Bay Books. 2005. ISBN 978-0-316-01066-5.

12. *The intuitive psychologist and his shortcomings: Distortions in the attribution process.* Lee Ross. In Berkowitz, L. Advances in experimental social psychology. New York: Academic Press. pp. 173–220. 1977. ISBN 0-12-015210-X.

13. *The Person and the Situation: Perspectives of Social Psychology.* Lee Ross, Richard E. Nisbett. Pinter & Martin Publishers. 2011.

14. *6-Hour Safety Culture: How to Sustainably Reduce Human Error and Risk, (and do what training alone can't do).* Tim Autrey. Human Performance Association. 2015. ISBN: 978-0996409810.

Chapter 4: What is a Just Culture?

1. *Managing the Risks of Organizational Accidents.* James Reason. Ashgate. 1997. ISBN 1840141050.

2. *Whack-a-Mole. The Price We Pay for Expecting Perfection.* David Marx. Your Side Studios. 2009.

3. *Human Error.* James Reason. Cambridge University Press. 1990. ISBN 0521314194.

4. *The Field Guide to Understanding Human Error.* Sidney Dekker. Ashgate Publishing. 2006. ISBN 978-0-7546-4826-0.

5. *Just Culture. Balancing Safety and Accountability*. Sidney Dekker. CRC Press. Second Edition. 2016. ISBN 1409440605.

6. *https://www.parents.com/baby/safety/car/youd-never-forget-your-child-in-the-car-right/*

7. *https://www.parents.com/parenting/better-parenting/advice/7-ways-to-not-forget-your-child-in-the-car/*

8. Hindsight Bias. *Perspectives on Psychological Science.* Neal J. Roese, Kathleen D. Vohs. Volume: 7 issue: 5, page(s): 411-426

Chapter 5: You Can't Coach Stupid

1. *Mindset. The New Psychology of Success.* Carol Dweck. Ballantine Books. 2008. ISBN 978-0-345-47232-8.

2. *Keen to Help? Managers' IPT and Their Subsequent Employee Coaching.* Heslin, VandeWalle, and Latham. Personnel Psychology 59 (2006) 871-902.

3. *http://fortune.com/2014/06/19/career-wise-is-it-better-to-be-smart-or-hardworking/*

Chapter 6: Your Dangerous Distracted Mind

1. End Distracted Driving. EndDD. *https://www.enddd.org/about-enddd/*

2. *http://www.txtresponsibly.org/*

3. *https://www.nhtsa.gov/risky-driving/distracted-driving*

4. *Momentary interruptions can derail the train of thought.* Altmann, E. M., Trafton, J. G., & Hambrick, D. Z. Journal of Experimental Psychology: General, 143*(1), 215-226.*

5. *A wandering mind is an unhappy mind.* Killingsworth, M. A., & Gilbert, D. T. (2010). Science, 330(6006), 932-932.

Chapter 7: I See What I Want to See

1. *Error Elimination Tools*. Practicing Perfection Institute, Inc. 2016. www.ppiweb.com

2. *Out of the Crisis.* W. Edwards Deming. Massachusetts Institute of Technology. Cambridge, MA. 1986.

Chapter 8: My Stuff is Worth More

1. *Thinking, Fast and Slow* by Daniel Kahneman Farrar, Straus and Giroux. 2011. ISBN-13: 978-0374275631

2. Kahneman, Daniel; Knetsch, Jack L.; Thaler, Richard H. (1990). *Experimental Tests of the Endowment Effect and the Coase Theorem.* Journal of Political Economy. 98 (6): 1325–1348.

3. *http://www.uxmatters.com/mt/archives/2011/03/how-anchoring-ordering-framing-and-loss-aversion-affect-decision-making.php*

4. *http://www.beinghuman.org/article/loss-aversion*

5. *The IKEA Effect: When Labor Leads to Love (Working Paper 11-091).* Michael I. Norton, Daniel Mochon, and Dan Ariely. Harvard Business School. 2011.

6. Biyalogorsky, Eyal, William Boulding, and Richard Staelin (2006), *Stuck in the Past: Why Managers Persist with New Product Failures*, Journal of Marketing, 70 (April), 108-21.

7. *http://leansixsigmadefinition.com/glossary/5s/*

Chapter 9: I'm Just Unlucky

1. John W. Jones and Lisa Wuebker. *Development and Validation of the Safety Locus of Control Scale.* Perceptual and Motor Skills. Vol 61, Issue 1, pp. 151 – 161.

2. Jones, J. W. *The Safety Locus of Control (SLC) Scale*. St. Paul, MN: The St. Paul Companies, 1983.

3. *Safety education and control: A tool to measure the safety locus of control.* Maryam Amidi Mazaheri, Alireza Hidarnia, and Fazlollah Ghofranipour. Journal of Education and Health Promotion. August 22, 2012.

4. J.B. Rotter. (1966). *Generalized expectancies for internal versus external control of reinforcement.* Psychological Monographs, 80, (1, Whole No. 609).

5. solutionsforresilience.com/locus-of-control

6. *Man's Search For Meaning.* Viktor Frankl. Beacon Press. 2006. ISBN-13: 978-0807014295

7. *The Locus of Control Construct's Various Means of Measurement: A researcher's guide to some of the more commonly used Locus of Control scales.* Russ Hill. Will To Power Press. 2011. ISBN 978-0-9833464-3-2

8. *Safety locus of control as a predictor of industrial accidents and injuries.* Lisa J. Wuebker. Journal of Business and Psychology, 1986, Volume 1, Number 1.

9. *Teach Internal Locus of Control.* Russ Hill. Will To Power Press. 2013. ISBN 978-0-9833464-0-1.

10. *Error communication in young farm workers: Its relationship to safety climate and safety locus of control.* Work and Stress. 23, 297-312.

Chapter 10: Mind Over Matter

1. Thomas K. Srull and Robert S. Wyer. *The Role of Category Accessibility in the Interpretation of Information About Persons: Some Determinants and Implications,* Journal of Personality and Social Psychology 37 (1979): 1660-1672.

2. John A. Bargh, Mark Chen, and Lara Burrows. *Automaticity of Social Behavior: Direct Effects of Trait Construct and Stereotype Activation on Action,* Journal of Personality and Social Psychology 71, no. 2 (1996): 230-244.

3. *Blink: The Power of Thinking Without Thinking.* Malcolm Gladwell. Back Bay Books. New York, NY. 2005. ISBN 978-0-316-17232-5 (hc).

4. *Predictably Irrational. The Hidden Forces That Shape Our Decisions.* Dan Ariely. Harper-Collins Publishers. New York. 2009. ISBN 978-0-06-135324-6.

5. Margaret Shih, Todd Pittinsky, and Nalinin Ambady. *Stereotype Susceptibility: Identity Salience and Shifts in Quantitative Performance.* Psychological Science. 1999.

6. SumiRiko in Bluffton, Ohio uses the Safety Pledge as part of their daily crew pre-shift meetings.

7. San Bolkan and Peter A. Andersen. *Image Induction and Social Influence: Explication and Initial Tests.* Basic and Applied Social Psychology 31(4):317-324 · November 2009.

8. *Pre-Suasion: A Revolutionary Way to Influence and Persuade.* Robert Cialdini. Simon and Schuster. New York, NY. 2016. ISBN 978-1-5011-0979-9.

Chapter 11: When Feelings Go Viral

1. *https://www.psychologytoday.com/blog/the-science-work/201410/faster-speeding-text-emotional-contagion-work*

2. *The Ripple Effect: Emotional Contagion and Its Influence on Group Behavior.* Sigale Barsade. Administrative Science Quarterly. Vol 47, No. 4. December 2002. pp 644 - 675.

3. *Well Said! Presentations and Conversations That Get Results.* Darlene Price. Amacom. 2012. ISBN-10: 0814417876

4. *Snap: Making the Most of First Impressions, Body Language, and Charisma.* Patti Wood. New World Library. 2012. ISBN-10: 1577319397

Chapter 12: The Power of Acknowledging

1. *Payoff - The Hidden Logic That Shapes Our Motivations.* Dan Ariely. Simon & Schuster. New York. 2016.

2. *Drive. The Surprising Truth About What Motivates Us.* Dan Pink. Riverhead Books. 2011. ISBN-10: 1594484805.

3. *The Ideal Praise-to-Criticism Ratio.* Jack Zenger and Joseph Folkman. Harvard Business Review. March 15, 2013.

4. *Can You Cope With Criticism at Work?* Vanessa Ko. CNN Business. April 14, 2013.

5. *The Magic Relationship Ratio, According to Science.* Kyle Benson. October 4, 2017. https://www.gottman.com/blog/the-magic-relationship-ratio-according-science/

6. *The Progress Principle. Using Small Wins to Ignite Joy, Engagement, and Creativity at Work.* Theresa Amabile and Steven Kramer. Harvard Business Review Press. Boston. 2011. ISBN: 978-1-4221-9857-5.

Chapter 13: You Can Do It!

1. Eden, Dov. *Leadership and Expectations: Pygmalion Effects and Other Self-Fulfilling Prophecies in Organizations.* Leadership Quarterly, 3(4), 271-305. (1992). JAI Press Inc.

2. Rosenthal, R., & Jacobson, L. (1968). *Pygmalion in the classroom: Teacher expectation and pupils' intellectual development.* New York: Holt, Reinhart & Winston.

3. Rosenthal, Robert; Jacobson, Lenore (1992). *Pygmalion in the classroom* (Expanded ed.). New York: Irvington.

4. *http://www.psychologytoday.com/blog/brain-babble/201412/why-were-so-easily-fooled-and-why-it-matters*

5.*http://psych.wisc.edu/braun/281/Intelligence/LabellingEffects.htm*

Chapter 14: Respected and Connected

1. Sinek, Simon. (2009). *Start With Why. How Great Leaders Inspire Everyone To Take Action.* Penguin Group. ISBN 978-1-59184-290-4.

2. Rogers, Everett M. (1983). *Diffusion of Innovations.* New York: Free Press. ISBN 978-0-02-926650-2

3. Gladwell, Malcolm (2002). *The Tipping Point: How Little Things Can Make a Big Difference.* Little, Brown, and Company. ISBN-13: 978-0-31634-662-7.

4. *Influencer. The Power to Change Anything.* Patterson, Grenny, Maxfield, McMillan, & Switzler. McGraw-Hill. 2008.

Chapter 15: If You Will, I Will

1. *Fellow airline passengers influence what you buy.* Stanford Business. February 6, 2015. http://stanford.io/1vwglQh

2. *Influencer. The Power to Change Anything.* Patterson, Grenny, Maxfield, McMillan, & Switzler. McGraw-Hill. 2008.

3. Asch, S. E. (1951). Effects of group pressure upon the modification and distortion of judgment. In H. Guetzkow (ed.) *Groups, leadership and men.* Pittsburgh, PA: Carnegie Press.

4. *https://www.simplypsychology.org/asch-conformity.html*

Crutchfield, R. (1955). *Conformity and Character.* American Psychologist, 10, 191-198.

5. Cialdini, R. B., Reno, R. R., & Kallgren, C. A. (1990). *A focus theory of normative conduct: Recycling the concept of norms to reduce littering in public places.* Journal of Personality and Social Psychology, 58, 1015–1026.

Chapter 16: Silence is Not Golden

1. Geert Hofstede. *Culture's Consequences: Comparing Values, Behaviors, Institutions, and Organizations Across Nations.* Sage Publications. Thousand Oaks, CA. 2001

2. Robert L. Helmreich and Ashleigh Merritt. "Culture in the Cockpit: Do Hofstede's Dimensions Replicate?" *Journal of Cross-Cultural Psychology* 31, no.3 (May 2000): 283-301.

3. "Culture May Play Role In Flight Safety -- Boeing Study Finds Higher Aviation Accident Rates Among Nations Where Individualism Not The Norm". Don Phillips. *The Washington Post.* August 22, 1994.

4. Ute Fischer and Judith Orasanu. *"Cultural Diversity and Crew Communication."* Astronautical Congress. Amsterdam. October 1999.

5. *One Simple Skill to Overcome Peer Pressure.* https://youtu.be/1-U6QTRTZSc

6. Asch, S. E. (1956). Studies of independence and conformity: I. A minority of one against a unanimous majority. Psychological monographs: General and applied, 70(9), 1-70.

7. Don Merrell. *"I Chose To Look The Other Way."* Powerlineman Magazine. Fall, 2002.

8. *Would You Watch Out For My Safety?* John Drebinger, Jr. http://www.drebinger.com/?page_id=774

Chapter 17: Show and Tell

1. *Long Walk to Freedom.* Nelson Mandela. Back Bay Books. 1995. ISBN-13: 978-0316548182

2. *Worst to First.* Gordon Bethune. Wiley Publishing. 1999. ISBN-13: 978-0471356523

3. *Managing the Flow of Technology.* Thomas J. Allen. MIT Press. 1984. ISBN-10: 0262510278

4. *Leaders Look for Symbolism.* Meg Whitman, HP. Interview by Tina Seelig of Stanford University. February, 2017.

5. *https://www.inc.com/jim-schleckser/use-stories-and-symbols-to-build-a-powerful-culture.html*

6. *Leading at the Edge. Leadership Lessons from the Extraordinary Saga of Shackleton's Antarctic Expedition.* Dennis N.T. Perkins. AMACOM. New York. 2000.

7. *South: A Memoir of the Endurance Voyage.* Ernest Shackleton. New York: Carroll & Graf. 1999.

Chapter 18: Faster and Safer

1. *Quick Changeover for Operators: The SMED System.* (Based on A Revolution in Manufacturing: The SMED System by Shigeo Shingo). Productivity Press. 1996. ISBN 978-1-56327-125-0.

2. *Inside Out - Rethinking Traditional Safety Management Paradigms.* Larry Wilson and Gary Higbee. Ecolab Limited. 2012. ISBN 978-0-9881577-0-5.

3. *Faster, Better, Cheaper in the History of Manufacturing: From the Stone Age to Lean Manufacturing and Beyond.* Christoph Roser. Productivity Press. 2016. ISBN-10: 9781498756303.

Leap Technologies. Rapid Action™
https://www.improvefaster.com/

Chapter 19: Why Should I Listen?

1. *Control or Caring? What is your motive for a safety conversation?* David A. Galloway. EHS Today. March 14, 2016.

2. *https://www.continuousmile.com/safety/safety-conversations-most-impact/*

Chapter 20: Are You Asking or Telling?

1. *Humble Inquiry.* Edgar H. Schein. Berrett-Koehler Publishers. San Francisco. 2013. ISBN 978-1-60994-981-5

2. *http://www.free-management-ebooks.com/faqcm/active-06.htm#sthash.v9eGHH9O.dpuf*

3. *https://www.continuousmile.com/communication/active-listening-hard-work/*

4. *https://www.forbes.com/sites/joefolkman/2015/09/08/ tell-ask-listen-the-3-steps-to-great-communication-as-a-leader/#7cefcfb33a0e*

Chapter 21: The Right Conversation

1. *Leadership is a Conversation.* Boris Groysberg and Michael Slind. Harvard Business Review. June, 2012.

2. *Error Elimination Tools™.* Practicing Perfection Institute, Inc. www.ppiweb.com 2015.

3. *Managing the Risks of Organizational Accidents.* Reason, James. Hampshire, England: Ashgate Publishing Limited. 1997.

4. *Whack-a-Mole. The Price We Pay For Expecting Perfection*, by David Marx. Your Side Studios. 2009.

5. *Out of the Crisis.* W. Edwards Deming. Massachusetts Institute of Technology. Cambridge, MA. 1982. ISBN 0-911379-01-0.

6. *Caring Enough to Confront* by David Augsburger. 3rd Edition. Revell. Scottdale, PA. 2009. ISBN-10: 0800724607.

9. *https://www.continuousmile.com/safety-conversation-guide/*

ACKNOWLEDGEMENTS

My earliest exposure to leadership came while I was a member of a Boy Scout troop in Pennsylvania. The Scoutmaster, Merle Eisenhart, had a powerful influence on the lives of hundreds of young boys who went through the program under his direction. Mr. Eisenhart treated all scouts with a tough kind of love that positively shaped their respective lives for years to come. Even today, I consider achieving the rank of Eagle Scout as one of my proudest accomplishments because of the effort required and the lessons I learned along the way.

This book was made possible by a great many people whom I have had the privilege to know in either a professional or personal capacity. As I recounted past experiences, it brought back memories of events that helped form my perspective of what it means to be an effective leader. Most of these experiences are linked to my career in the paper industry. I am grateful to have worked with men and women who exemplified many of the traits embodied in this book.

I would like to personally express my appreciation to the colleagues who provided feedback and comments on the manuscript. This is a much better book because of their suggestions and critique. Thanks to: Allan Bohn, David Bonistall, Daniel Clark, Kathy Collins, Neil Cronkhite, Scott Grimes, Tracey Hotopp, Mark Lukacs, Michael Pecoraro, Brian R. Smith, and Sean Wallace.

Finally I want to thank my wife, Leesa. She was my primary editor. She carefully read every word numerous times and returned the marked up manuscript with a smile and a kiss. I'm so lucky she is my partner for life.

ABOUT THE AUTHOR

After graduating from Penn State, David Galloway started his career in the paper industry. Over the next 35 years, he gained experience and held leadership roles in process engineering, operations, research, product development, quality, logistics, and strategy.

As a Certified Master Black Belt, Certified Master Facilitator, and Lean Six Sigma Deployment Director, David has applied his knowledge in leadership, change management, human performance, and Lean Six Sigma to improve organizational performance.

David has a passion for workplace safety and the psychology of personal risk taking. He helps clients develop an effective safety strategy with an emphasis on leadership. David is the author of a safety culture assessment tool that enables organizations to place themselves on *The Safety Leadership Continuum*™.

David facilitates a workshop for supervisors designed to strengthen safety conversation and leadership skills. Many of the key concepts in this book are included in the agenda.

He also developed a key tool for managers and supervisors: *Safety Conversation Guide,* which is available as a mobile app for Apple and Android users. Using this tool fosters a culture of commitment by promoting trust, learning, and improvement through sincere and effective personal safety conversations. The *Pro* version of the *Safety Conversation Guide* gives clients the ability to quickly and easily document individual and group safety conversations. It is a subscription-based service designed for organizations that want to track these conversations, improve employee engagement, and strengthen their safety culture.

David is a keynote speaker who has authored over 50 articles about safety and safety leadership. He is the Founder and President of Continuous MILE Consulting, LLC. For more information visit *www.ContinuousMile.com*

David lives in Springboro, Ohio with his wife, Leesa. They have two children and two grandchildren.

Made in United States
Troutdale, OR
08/28/2023

12404847R00156